PROBABILITY
THEORY

BY

A. M. ARTHURS

LONDON: Routledge & Kegan Paul Ltd
NEW YORK: Dover Publications Inc.

First published 1965
in Great Britain by
Routledge & Kegan Paul Ltd
Broadway House, 68–74 Carter Lane
London, E.C.4
and in the U.S.A. by
Dover Publications Inc.
180 Varick Street
New York, 10014
Second impression 1967
Third impression 1969
Printed in Great Britain by Photolithography
Unwin Brothers Limited, Woking and London
SBN 7100 4359 7

Preface

THE PURPOSE of this book is to give an elementary introduction to the mathematical theory of probability. This theory was founded in the seventeenth century by Pascal and Fermat for the solution of problems connected with games of chance. Modern developments, however, have produced a theory with an enormous variety of applications and one which appears to lie at the very foundations of scientific thought.

In all probabilistic reasoning there is a basic framework consisting of a set called the sample space, and subsets called simple events, which have probabilities assigned to them. The central aim of the text is to describe this framework as simply and clearly as possible and to develop the theory on axiomatic lines. Abstract ideas are motivated by examples of a practical nature since the interplay between theory and application is especially strong in this branch of mathematics.

Each chapter ends with a set of exercises. Some of these exercises first appeared in the books listed on page 76 and my thanks are due to John Wiley and Sons, New York, for permission to use them.

I am grateful to Professor P. B. Kennedy and Professor W. Ledermann for many invaluable suggestions during the preparation of this book.

A. M. ARTHURS

University of York

v

Contents

Contents

CHAPTER ONE

Set Theory

1.1. Introduction

The mathematical theory of probability, like any other branch of mathematics, deals with certain objects that possess a mathematical structure. Before we can describe what probability is about, we must understand the underlying ideas about collections of objects, that is, we must know some elementary set theory*.

1.2. Sets and subsets

The notion of set is one of the most primitive in mathematics — more primitive, for example, than the notion of number. We cannot expect a definition of such a primitive notion, but we can give it the following:

Description. A *set* is a collection of objects thought of as a whole.

The objects, of which the set is a collection, are called *elements* or *members* of the set.

If A is a set, and x is an element of A, we write

$$x \in A. \tag{1}$$

If x does *not* belong to A we write

$$x \notin A. \tag{2}$$

* For a detailed treatment of set theory, see S. Swierczkowski, *Sets and Numbers*, Routledge & Kegan Paul, Library of Mathematics.

We may be able to specify a set by writing down names for all its elements. For instance, if $A_1, ..., A_n$ are objects, the set consisting of precisely these objects will be written

$$\{A_1, A_2, ..., A_n\}. \tag{3}$$

We remark here that there is a distinction between an object A_1 and the set $\{A_1\}$ consisting of that object alone. Thus

$$A_1 \in \{A_1\} \tag{4}$$

is a true statement, whereas

$$A_1 \in A_1 \tag{5}$$

is a false statement, because a set cannot be its own element.

It is convenient to introduce the notion of a set that has no members, the so-called *null* or *empty* set $\varnothing$. The set $\varnothing$ is such that the statement

$$x \in \varnothing$$

is false whatever the object x is.

Sometimes we cannot list all the elements of a set, for example, the set of all positive integers. To cover cases like this we introduce for a set A the notation

$$A = \{x; \quad \}, \tag{6}$$

where we write in the blank space the property that 'puts' x in A. The notation is to be read 'the set of x such that'. For example, the set of all positive integers is written

$$\{x; x \text{ is a positive integer}\}.$$

A basic relation between sets is *inclusion* of one set in another.

Definition. Subset. $A \supset B$ or $B \subset A$ (read 'A includes or contains B' or 'B is a *subset* of A') means that the set B consists of elements of the set A. That is,

$$\text{if } x \in B \text{ then } x \in A. \tag{7}$$

2

In accordance with this definition of subset, a set A is always a subset of itself. It is also true that the empty set is a subset of A. If $B \subset A$ and $B \neq A$, then B is called a *proper* subset of A.

Two sets are identical if they have precisely the same elements. Thus, if A and B are sets, then $A = B$ if and only if both

$$A \subset B \text{ and } B \subset A. \tag{8}$$

Frequent use is made of this fact in proving the equality of two sets.

In a given context there is a 'large set' of which all the sets we talk about are subsets. This large set is usually called the *space* or *universal set* and denoted by S.

Venn diagrams. Sets and subsets can be represented in a pictorial way by means of Venn diagrams. The rectangle S in the diagram represents the space S, and the *elements* of S are represented by points in the rectangle. Sets of elements of S, such as A and B, are represented by the points inside closed curves.

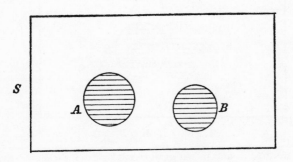

Fig. 1.1.   Venn diagram

3

Subsets are represented in a Venn diagram as follows

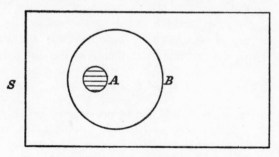

FIG. 1.2. *A* is a subset of *B* (*A* ⊂ *B*)

1.3. Set operations

Definition. Union. By the *union*, *A* ∪ *B*, of two sets *A* and *B* we mean the set of elements which belong to at least one of the two sets *A*, *B*; that is

$$A \cup B = \{x; x \in A \text{ or } x \in B\}.* \qquad (9)$$

The union of *A* and *B* is indicated by the shaded area in Fig. 1.3.

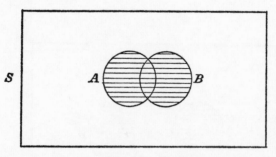

FIG. 1.3. *A* ∪ *B*

---

* The word 'or' is used in mathematics and logic in the inclusive sense. Thus, the statement '*P* or *Q*' is the mathematical expression for 'either *P* or *Q* or both'.

Set operations

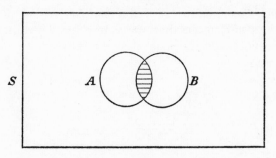

Fig. 1.4. $A \cap B$

Definition. Intersection. By the *intersection*, $A \cap B$, of two sets A and B we mean the set of elements which belong to both A and B; that is

$$A \cap B = \{x; x \in A \text{ and } x \in B\}. \tag{10}$$

The intersection of A and B is indicated by the shaded area in Fig. 1.4.

Definition. Disjoint, mutually exclusive. Two sets A and B are said to be *disjoint*, or *mutually exclusive*, if they have no elements in common, that is

$$A \cap B = \varnothing. \tag{11}$$

Disjoint sets are illustrated in Fig. 1.5.

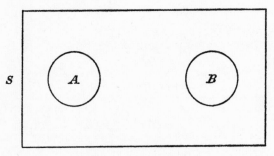

Fig. 1.5. Disjoint sets

Definition. Complement. Let $A \subset S$. The *complement* of A in S is the set of elements that belong to S but not to A. The complement of A in S is denoted by A'. Thus

$$A' = \{x; x \notin A\}. \tag{12}$$

The shaded part of Fig. 1.6 indicates the complement of A.

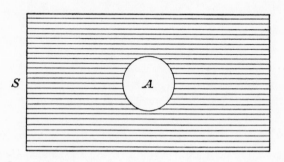

Fig. 1.6. A'

Note that $(A')' = A$ for any set A, and that, if S is the space, then $S' = \varnothing$ and $\varnothing' = S$.

The symbols $\cup$ and $\cap$ are operations between any two sets just as $+$ and $\times$ are operations between any two numbers, and they obey similar laws:

Commutative law: $A \cup B = B \cup A$,
$\qquad\qquad\qquad\; A \cap B = B \cap A.$ $\tag{13}$

Associative law: $(A \cup B) \cup C = A \cup (B \cup C)$,
$\qquad\qquad\quad\; (A \cap B) \cap C = A \cap (B \cap C).$ $\tag{14}$

Distributive law: $A \cap (B \cup C) = (A \cap B) \cup (A \cap C)$,
$\qquad\qquad\quad A \cup (B \cap C) = (A \cup B) \cap (A \cup C).$
$\tag{15}$

Idempotency law: $A \cup A = A, A \cap A = A.$ $\tag{16}$

Finally we note two very useful identities

$$(A \cup B)' = A' \cap B', (A \cap B)' = A' \cup B', \qquad (17)$$

which are sometimes called *de Morgan's laws*.

All of these laws are readily illustrated in Venn diagrams. Rigorous proofs can be found for example in Swierczkowski's book.

EXERCISE FOR CHAPTER ONE

1. Let $A \subset S$. Prove that

    (a) $A \cup A' = S$.   (d) $A \cap S = A$.

    (b) $A \cap A' = \varnothing$.   (e) $A \cup \varnothing = A$.

    (c) $A \cup S = S$.   (f) $A \cap \varnothing = \varnothing$.

CHAPTER TWO

Probability and Discrete Sample Spaces

2.1. Introduction

In Chapter One we outlined the basic ideas of set theory. In this chapter we shall show how the abstract notion of sets may be used in the development of the mathematical theory of probability.

Most people know that probability is somehow connected with 'chance'. Indeed, the theory of probability was founded by Pascal and Fermat to describe certain games of chance and to calculate various probabilities. For instance, in n tosses of a coin, one might want to know the chance of getting m heads. However, probability theory has a much wider field of application. For example, the theory is used to study problems in economics, genetics, sociology, astronomy and physics.

With many diverse applications in mind, we must develop the theory in abstract form, so that it is not tied down to one particular kind of problem. Now we might ask how a mathematical theory is developed. Consider, for instance, the case of geometry. There one starts with undefined concepts like 'point' and 'straight line'. Next, certain statements are made about the properties of these concepts and the relations between them. For example, one may state that 'two points determine a line'. These statements are called the rules or axioms of the theory. The important thing about these axioms is that they are chosen by us and are not decided for us by some outside factors. From the axioms of a theory various theorems can be logically deduced. Such a procedure leads to a mathematical theory which may or may not be useful in the study of real processes.

The abstract concepts of geometry are in practice identified with certain physical objects. For instance, the abstract notion of a 'point' is often used in mechanics to describe a physical body like the sun. In general, if there is a rule by which we identify abstract concepts with physical objects, then we have what is called a *mathematical model* of the real process, and different models can describe the same empirical situation.

In this book we are concerned with the study of *probability models*. The theory to be developed, however, is limited to one particular aspect of 'chance'. Judgments such as 'it will probably be fine tomorrow' are of interest to the logician, but they do not concern us. Rather, our probabilities refer to the 'possible outcomes of certain experiments or observations'. The probability model will be seen to involve two things:

(i) choosing a set to represent the possible outcomes,
and   (ii) allocating probabilities to these possible outcomes.
Once we have specified a probability model by (i) and (ii), we shall apply the general theorems derived in this chapter and develop the theoretical consequences.

2.2. Sample spaces and events

Any mathematical model involves idealization, and our first idealization concerns the possible outcomes of an 'experiment' or 'observation'. If we want to construct an abstract model, we must decide what constitutes a possible outcome of the (idealized) experiment. To illustrate this, consider the 'experiment' of tossing two coins. How do we list the possible outcomes of this experiment? It may be done in a number of ways, depending on what we are interested in. For example, we may be interested in whether each coin falls heads (H) or tails (T). Then the possible outcomes are

$$(H,H) \quad (H,T) \quad (T,H) \quad (T,T) \tag{1}$$

Every outcome of the experiment corresponds to exactly one member of the set (1). In this model we are deliberately omitting the possibility that when the coins fall they might roll away or stand on their edges.

On the other hand, we may be interested only in whether the coins fall alike (A) or different (D). Then we could list the possible outcomes as

$$(A)\,(D), \tag{2}$$

giving, on allocating probabilities, a second model of the experiment.

The possible outcomes define the idealized experiment and they are usually called *sample points*.

The collection of all sample points is called the *sample space S* of the model.

The notion of *event* can now be introduced. An event is a *subset* of *S*. For example, consider the sample space given by equation (1). If A is the subset

$$A = \{(H,H),(H,T),(T,H)\}, \tag{3}$$

and B is the subset

$$B = \{(H,H)\}, \tag{4}$$

then we can say that A is the event that there is at least one head, and B is the event that there are two heads. Thus an event may contain one or more sample points. An event, such as B, which contains only *one* sample point, is called a *simple event*.

With any two events A and B we can associate two new events defined by the conditions 'either A or B or both occur' and 'both A and B occur'. In the notation of set theory these events are written $A \cup B$ and $A \cap B$, respectively. The event $A \cup B$, the union of A and B, means that at least one of the events A and B occurs; it contains all sample points except

10

those that belong neither to A nor to B. The event $A \cap B$, the intersection of A and B, contains all sample points which are common to A and B. In probability theory we can describe the event $A \cap B$ as the simultaneous occurrence of A and B. If there are no points common to A and B, then

$$A \cap B = \emptyset, \qquad (5)$$

and we say that the events A and B are *mutually exclusive*.

For any event $A \subset S$ we also introduce the *complementary event* A' which contains the sample points of S not in A. Thus A' is the event that A does not occur.

To illustrate some of these ideas we consider the sample space S given by the set of three points

$$S = \{e_1, e_2, e_3\}. \qquad (6)$$

The non-empty subsets of S are

$$\{e_1\}, \{e_1, e_2\},$$
$$\{e_2\}, \{e_1, e_3\}, \{e_1, e_2, e_3\},$$
$$\{e_3\}, \{e_2, e_3\}. \qquad (7)$$

Each of these subsets is an event. The empty set is also an event, though trivial. The subsets

$$E_1 = \{e_1\}, E_2 = \{e_2\}, E_3 = \{e_3\}$$

contain just one sample point each and hence are simple events. It is clear from their definition that simple events are mutually exclusive. For instance, $E_1 \cap E_2 = \emptyset$. Every event, apart from the empty set, is the union of one or more distinct simple events. For example, $E_2 \cup E_3 = \{e_2, e_3\}$, and we know that this is an event since it is a subset of S. Also, the whole sample space S can be written as the union of all distinct simple events since

$$S = \{e_1, e_2, e_3\} = E_1 \cup E_2 \cup E_3.$$

We have thus dealt with one aspect of a probability model, namely, the setting up of a sample space S of all possible outcomes.

In the next section we consider the second important aspect of a probability model. This is the allocation of probabilities to the elements of a sample space.

2.3. Probabilities in discrete sample spaces

We shall confine our attention here to sample spaces which are *discrete*, that is, they contain only a finite number of sample points or an infinite number of points which can be arranged in a simple sequence $e_1, e_2, \ldots$. Thus we are excluding for the moment sample spaces such as the points on the real line.

Let the sample space S be the set

$$S = \{e_1, e_2, \ldots\} = E_1 \cup E_2 \cup \ldots, \tag{8}$$

where $E_i = \{e_i\}$ are the simple events in S. Then we assume that to each event E in S we can assign a non-negative real number $P[E]$, called the *probability* of E. This number satisfies the following axioms:

Axioms for probability

    I $P[E] \geqslant 0$ for every event E.

    II $P[S] = 1$ for the certain event S.

    III The probability $P[A]$ of any event A is the sum of the probabilities of the simple events whose union is A.

If the sample space S is the union of the distinct simple events $E_1, E_2, \ldots$, it follows from axioms II and III that

$$P[S] = P[E_1] + P[E_2] + \ldots = 1. \tag{9}$$

An important consequence of axiom III is that if E and F are mutually exclusive events, so that $E \cap F = \varnothing$, then

$$P[E \cup F] = P[E] + P[F]. \tag{10}$$

Theorem 1. $$P[\varnothing] = 0, \tag{11}$$

where $\varnothing$ is the empty set.

Proof. In equation (10) take $E = S$ and $F = \varnothing$, which are mutually exclusive. Then

$$P[S \cup \varnothing] = P[S] + P[\varnothing]. \tag{12}$$

Now $$S \cup \varnothing = S \tag{13}$$

and so $$P[S \cup \varnothing] = P[S]. \tag{14}$$

Subtracting (14) from (12) we find that

$$P[\varnothing] = 0$$

as required.

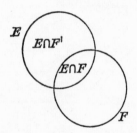

Theorem 2. $$P[E \cap F'] = P[E] - P[E \cap F]. \tag{15}$$

Proof. The events $E \cap F$ and $E \cap F'$ are disjoint and their union is E (see diagram).

Thus, $$P[E] = P[(E \cap F) \cup (E \cap F')]$$
$$= P[E \cap F] + P[E \cap F'] \text{ by equation (10).}$$

Therefore, $$P[E \cap F'] = P[E] - P[E \cap F].$$

Theorem 3. Complementary events.

$$P[F'] = 1 - P[F]. \tag{16}$$

Proof. Take $E = S$ in theorem 2.

Since $\quad S \cap F' = F', S \cap F = F,$

we have $\quad P[F'] = P[S] - P[F]$ from theorem 2,

$$= 1 - P[F] \text{ by axiom II.}$$

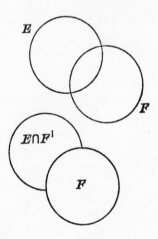

Theorem 4. Probability of E or F.*

$$P[E \cup F] = P[E] + P[F] - P[E \cap F]. \qquad (17)$$

Proof. Write $E \cup F$ as the union of two disjoint events $E \cap F'$ and F (see diagram). Then

$$P[E \cup F] = P[(E \cap F') \cup F]$$
$$= P[E \cap F'] + P[F] \text{ by equation (10)},$$
$$= P[E] - P[E \cap F] + P[F] \text{ by theorem 2,}$$

which is the required result. It is a generalisation of the result stated in equation (10).

---

* See footnote, page 4, for mathematical meaning of 'or'.

2.4. Equally likely outcomes

A large part of probability theory is based on the idea of
'equally likely outcomes', that is, outcomes that have the same
chance of occurring. For instance, if a coin is tossed it seems
reasonable to suppose that the coin is just as likely to fall
'heads' as to fall 'tails'. So the two possible outcomes, heads
and tails, are considered to be equally likely. The probability
model for this experiment is then

$$\text{sample space } S = \{H,T\}, \tag{18}$$

$$\text{with } P[\{H\}] = P[\{T\}] = \tfrac{1}{2}. \tag{19}$$

We might describe this model intuitively by saying that if we
imagine the coin is tossed a great many times, then we would
expect 'heads', in the long run, to occur fifty times out of a
hundred.

Consider now a sample space S containing $N[S]$ simple
events. If we assume that these simple events are *equally likely*,
the probability of any event A defined on S takes a very simple
form. Let the possible simple events be $E_1, E_2, ..., E_n$ with
$n = N[S]$, and suppose that the union of exactly $N[A]$ of these
simple events corresponds to the event A. From axiom III
we have

$$P[A] = P[E_1] + P[E_2] + ... + P[E_k], \text{ with } k = N[A] \text{ here.}$$

But, for equally likely outcomes, we have from equation (9),

$$P[E_i] = \frac{1}{N[S]}, \quad i = 1, 2, ..., N[S].$$

Therefore $\qquad P[A] = \dfrac{N[A]}{N[S]}. \tag{20}$

Examples

(a) An ordinary six-sided die has the faces 1, 2, 3 and 4 coloured
white and the faces 5 and 6 coloured black. Let us find the probability that,
when the die is thrown once, the top face is (i) white, (ii) black.

Take the sample space

$$S = \{1,2,3,4,5,6\}, \tag{21}$$

and assume equally likely outcomes. Let A be the event that the top face is white, so that

$$A = \{1,2,3,4\} \tag{22}$$

and let B be the event that the top face is black,

$$B = \{5,6\}. \tag{23}$$

Then from equation (20)

$$P[A] = \frac{N[A]}{N[S]} = \frac{4}{6}, \tag{24}$$

and

$$P[B] = \frac{N[B]}{N[S]} = \frac{2}{6}. \tag{25}$$

We note that in this model the two outcomes 'white' and 'black' are not equally likely.

(b) A coin is tossed twice. Events E and F are defined as

$$E = \text{heads on first toss},$$

$$F = \text{heads on second toss}.$$

Let us find the probability of $E \cup F$. Take the sample space of four points

$$S = \{HH, HT, TH, TT\}, \tag{26}$$

and assume equally likely outcomes. Then

$$E = \{HH, HT\} \text{ and } F = \{HH, TH\},$$

giving

$$E \cup F = \{HH, HT, TH\}.$$

Therefore

$$P[E \cup F] = \frac{N[E \cup F]}{N[S]} = \frac{3}{4}. \tag{27}$$

We can check this result by using theorem 4:

$$P[E \cup F] = P[E] + P[F] - P[E \cap F].$$

Now

$$E \cap F = \{HH\},$$

and so

$$P[E \cap F] = \frac{N[E \cap F]}{N[S]} = \frac{1}{4}.$$

Also

$$P[E] = \frac{N[E]}{N[S]} = \frac{2}{4}, \; P[F] = \frac{N[F]}{N[S]} = \frac{2}{4}.$$

Hence

$$P[E \cup F] = \frac{2}{4} + \frac{2}{4} - \frac{1}{4} = \frac{3}{4}.$$

(c) Placing two objects in two cells. Let a and b be the two objects, and suppose that the cells are distinguishable. We consider two cases:

    I objects distinguishable,

    II objects indistinguishable.

The possible outcomes of this experiment can be described as follows:

I	II
$E_1 \{ab|-\}$	$F_1 \{* \; *|-\}$
$E_2 \{-|ab\}$	$F_2 \{-|* \; *\}$
$E_3 \{a|b\}$	
$E_4 \{b|a\}$	$F_3 \{*|*\}$

Each of these arrangements represents a simple event. In case I we suppose the possible outcomes are equally likely, so that

$$P[E_1] = P[E_2] = P[E_3] = P[E_4] = \frac{1}{4}. \tag{28}$$

In case II there are three outcomes, though we might argue that event F_3 occurs in two ways with identical results. This suggests, on the basis of case I, that we take

$$P[F_1] = P[F_2] = \frac{1}{4}, \; P[F_3] = \frac{1}{2}. \tag{29}$$

On the other hand, we could simply take the three outcomes to be equally likely,

$$P[F_1] = P[F_2] = P[F_3] = \frac{1}{3}, \tag{30}$$

and so obtain a second model for case II.

This example shows that different assignments of probabilities are compatible with the same sample space.

We might ask which model in case II is the more realistic? The answer is that each model is realistic under certain circumstances. For example, the assumptions which lead to equation (30) have been found to be valid for the statistical mechanics of photons and certain other particles. This remarkable fact was discovered by Bose and Einstein.

2.5. Conditional probability

Consider the sample space $S = \{1,2,3,4,5,6\}$ and let probabilities be assigned to the simple events. Let event $A = \{1,2\}$ and event $B = \{2,4,6\}$. We want to introduce the notion of conditional probability, which can be used to answer such questions as: what is the probability of B when A is known to have occurred? For the above case, if we know that x is less than 3 (event A), what can we say about the probability that x is even (event B)? The adjective 'conditional' is necessary in questions like this because we are evaluating probabilities subject to certain preassigned conditions.

To help us find the answer, consider a third event $C = \{2\}$ and ask for the conditional probability of B, given that C has already occurred. The intuitive answer is clear in this case. If x is known to be 2, then x is certainly even, and the probability must be 1. What made the answer easy was the fact that C was contained in B. The general question of conditional probability asks us to evaluate the extent to which the given event A is contained in the event B. Now the extent to which A is contained in B can be measured by the extent to which B and A are likely to occur simultaneously, that is, by $P[B \cap A]$. So the conditional probability of B, given that A has occurred, written as $P[B|A]$, is proportional to $P[B \cap A]$, and we may take

$$P[B|A] = k\, P[B \cap A], \tag{31}$$

with k some factor to be determined. We find k by noting that the conditional probability of A, given that A has occurred, is clearly 1. Thus

$$P[A|A] = k\, P[A \cap A] = k\, P[A] = 1,$$

giving

$$k = \frac{1}{P[A]}. \tag{32}$$

18

We are therefore led to state

Definition. The *conditional probability* of B, given A, is defined to be

$$P[B|A] = \frac{P[B \cap A]}{P[A]}, \tag{33}$$

provided that $P[A] \neq 0$. Conditional probabilities remain undefined when $P[A] = 0$.

For $B = \{2,4,6\}$ and $C = \{2\}$, the definition (33) gives the answer derived above, $P[B|C] = 1$. For $B = \{2,4,6\}$ and $A = \{1,2\}$ we obtain the reasonable figure $P[B|A] = \frac{1}{2}$, assuming equally likely outcomes; that is, if it is known that x is either 1 or 2, then x is even (event B) or odd (event $B' = $ complement of B) each with probability $\frac{1}{2}$.

Let $S = E_1 \cup E_2 \cup ... \cup E_n$ be a sample space consisting of n simple events E_i. If A is any non-empty proper subset of S, it can be shown that

$$\sum_{i=1}^{n} P[E_i|A] = 1. \tag{34}$$

This is analogous to the sum rule (9)

$$P[S] = P[E_1] + P[E_2] + ... + P[E_n] = 1$$

satisfied by ordinary probabilities.

Further, all general theorems on probabilities are valid for conditional probabilities. For example, theorem 4 becomes

$$P[A \cup B|C] = P[A|C] + P[B|C] - P[A \cap B|C]. \tag{35}$$

Also, setting $A = S$ in (33) we see that

$$P[B|S] = P[B], \tag{36}$$

which shows that ordinary probability is a special case of conditional probability.

Example

Throwing a die. Let the sample space be $S = \{1,2,3,4,5,6\}$ and assume equally likely outcomes. Consider events $A = \{1,2,3,4\}$ and $B = \{3,5\}$. Then $B \cap A = \{3\}$, and $P[B \cap A] = \frac{1}{6}$, $P[A] = \frac{4}{6}$. Hence the conditional probability of B, given A, is

$$P[B|A] = \frac{P[B \cap A]}{P[A]} = \frac{1}{4}.$$

We now describe a result which is mathematically equivalent to (33). It is known as

Bayes' Theorem. Let $S = E_1 \cup E_2 \cup \ldots \cup E_n$ be a sample space consisting of n simple events E_i, and let A be any non-empty subset of S. Then for each integer k $(1 \leqslant k \leqslant n)$,

$$P[E_k|A] = \frac{P[A|E_k]P[E_k]}{\sum\limits_{j=1}^{n} P[A|E_j]P[E_j]}. \qquad (37)$$

Proof. Let the simple events E_i be labelled so that

$$A = E_1 \cup E_2 \cup \ldots \cup E_m, \ 1 \leqslant m \leqslant n. \qquad (38)$$

Then
$$P[A] = \sum_{j=1}^{m} P[E_j].$$

Now
$$\sum_{j=1}^{n} P[A|E_j]P[E_j] = \sum_{j=1}^{n} P[A \cap E_j] \text{ from (33)},$$

$$= \sum_{j=1}^{m} P[E_j] \text{ using (38)},$$

$$= P[A].$$

So the right hand side of (37) is

$$\frac{P[A|E_k]P[E_k]}{P[A]} = \frac{P[A \cap E_k]}{P[A]}$$

$$= P[E_k|A],$$

which is the required result.

Bayes' Theorem relates the conditional probability of E_k, given A, to the conditional probabilities of A, given E_j and the probabilities of the E_j themselves. If the events E_j are called 'causes' then (37) can be regarded as a formula for the probability that the event A, which has occurred, is the result of the cause E_k.

2.6. Independent events

Definition. The events A and B are said to be *independent* if and only if

$$P[A \cap B] = P[A]P[B]. \tag{39}$$

Using the definition of conditional probability (33), we see that (39) implies that

$$P[B|A] = P[B] \text{ and } P[A|B] = P[A]. \tag{40}$$

If two events A and B do *not* satisfy equation (39), the events are *dependent*.

Theorem 5. If the events A and B are independent, then events A and B' are independent.

Proof. $P[A \cap B'] = P[A] - P[A \cap B]$ by theorem 2,

$\qquad\qquad = P[A] - P[A]P[B]$, since A and B are independent,

$\qquad\qquad = P[A] \{1 - P[B]\}$

$\qquad\qquad = P[A] P[B']$ by theorem 3.

Hence A and B' are independent.

Theorem 6. If A and B are independent events with non-zero probabilities, then the events A and B have a common sample point.

Proof. Let $\varnothing$ be the empty set. Then either $A \cap B = \varnothing$ or

$A \cap B \neq \varnothing$. If $A \cap B = \varnothing$, $P[A \cap B] = 0$. But $P[A \cap B] = P[A] P[B]$, since A and B are independent. So, either $P[A] = 0$ or $P[B] = 0$. This contradicts the hypothesis. Hence $A \cap B \neq \varnothing$, which proves the theorem.

Example 1

Independent events. Two true dice are thrown. We suppose the dice are distinguishable, and take the sample space represented by the 36 sample points

$$
S: \quad
\begin{array}{cccccc}
(1,1) & (1,2) & (1,3) & (1,4) & (1,5) & (1,6) \\
(2,1) & (2,2) & (2,3) & (2,4) & (2,5) & (2,6) \\
(3,1) & (3,2) & (3,3) & (3,4) & (3,5) & (3,6) \\
(4,1) & (4,2) & (4,3) & (4,4) & (4,5) & (4,6) \\
(5,1) & (5,2) & (5,3) & (5,4) & (5,5) & (5,6) \\
(6,1) & (6,2) & (6,3) & (6,4) & (6,5) & (6,6)
\end{array}
\qquad (41)
$$

We assume equally likely outcomes. Now consider the events $A = 1$ with first die, and $B =$ even number with second die.
From (41) we see that

$$N[A] = 6, \ N[B] = 18, \ N[A \cap B] = 3, \ N[S] = 36,$$

giving

$$P[A] = \frac{1}{6}, \ P[B] = \frac{1}{2}, \ P[A \cap B] = \frac{1}{12}, \text{ and so}$$

$$P[A \cap B] = P[A] P[B].$$

The events A and B are therefore independent.

Example 2

Dependent events. Let A and B be the following events of the experiment in example 1 above:

$$A = \{m + n = 11\},$$

$$B = \{n \neq 5\},$$

where (m,n) is a typical sample point. Then from (41)

$$P[A] = \frac{2}{36}, \ P[B] = \frac{30}{36}, \text{ and } P[A \cap B] = \frac{1}{36},$$

giving

$$P[A \cap B] \neq P[A] P[B].$$

Thus, A and B are dependent events.

To end this chapter we stress the difference between mutually exclusive events and independent events. Mutually exclusive events in general are *not* independent, while we know from theorem 6 that independent events in general are *not* mutually exclusive.

EXERCISES FOR CHAPTER TWO

1. Let A and B be arbitrary events defined on a sample space S. Find expressions in terms of subsets of S for the events that of A and B

(a) exactly none occurs,
(f) at least two occur,

(b) exactly one occurs,
(g) no more than none occurs,

(c) exactly two occur,
(h) no more than one occurs,

(d) at least none occurs,
(i) no more than two occur,

(e) at least one occurs,
(j) A occurs and B does not occur.

2. Let $S = \{1,2,3,4,5,6,7,8,9,10\}$, $A = \{1,2,3,4,5,6\}$ and $B = \{4,5,6,7,8,9\}$. For each of the events described in exercise 1, write out the numbers that are members of the event.

3. A and B are any two events defined on a sample space S. Show that

$$P[(A \cap B') \cup (B \cap A')] = P[A] + P[B] - 2P[A \cap B].$$

4. A, B and C are any three events defined on a sample space S. Show that

$$P[A \cup B \cup C] = P[A] + P[B] + P[C] - P[A \cap B] - P[B \cap C]$$
$$- P[C \cap A] + P[A \cap B \cap C].$$

5. Two dice are thrown. Let A be the event that the sum of the upper face numbers is odd, and B the event of at least one ace. Assuming a sample space of 36 points, list the sample points which belong to the events $A \cap B$, $A \cup B$, and $A \cap B'$. Find the probabilities of these events, assuming equally likely outcomes. (Feller)

6. A die is thrown as long as necessary for a six to turn up. Assuming that the six does not turn up at the first throw, find the probability that more than five throws will be necessary.

7. A die is loaded in such a manner that for $k = 1, ..., 6$, the probability of the face marked k turning up when the die is tossed is proportional to k. Find the probability that the outcome of a toss of the die will be an even number. (Parzen)

23

8. Let A and B be two independent events such that the probability is $\frac{1}{8}$ that they will occur simultaneously and $\frac{3}{8}$ that neither of them will occur. Find $P[A]$ and $P[B]$. (Parzen)

9. Let S be a sample space and suppose $A \subset S$, $B \subset S$. If $P[A] > 0$, show that

(i) $B \subset A$ implies $P[B|A] = \dfrac{P[B]}{P[A]}$,

(ii) $A \subset B$ implies $P[B|A] = 1$.

10. Let $S = E_1 \cup E_2 \cup \ldots \cup E_n$ be a sample space consisting of n simple events E_i. If A is any non-empty proper subset of S, show that

$$\sum_{i=1}^{n} P[E_i|A] = 1.$$

11. Two coins are tossed. What is the conditional probability that two heads result, given that there is at least one head?

12. Problem of points. On each play of a game, one of two players scores a point, and the two players have equal chances of gaining the point. Three points are required to win. If the players must leave the game when one has 2 points and the other has 1 point, how should the stakes be divided between them?

CHAPTER THREE

Sample Spaces with many Elements

3.1. Introduction

In the study of simple games of chance we are usually dealing with finite sample spaces in which the same probability is attributed to all sample points. For these cases, to compute the probability of an event A we then have to divide the number of sample points in A by the total number of sample points in the space (see equation (20) of chapter 2). But certain experiments, such as dealing a hand of bridge cards, contain a *very large* number of sample points and it is very inconvenient to list them all. So we require some method of *calculating* the number of sample points.

3.2. Permutations and combinations

We first of all wish to generalize the well-known fact that playing cards with 4 different suits and 13 face cards have $4 \times 13 = 52$ combinations or cards altogether.

Suppose we have two sets

$$S_1 = \{a_1, ..., a_m\} \text{ containing } m \text{ elements,}$$

and

$$S_2 = \{b_1, ..., b_n\} \text{ containing } n \text{ elements.}$$

Now choose one element, say x, from S_1 and one element y from S_2. The elements x and y make up what we shall call a sample (x,y). How many distinct samples (x,y) can be chosen?

We can answer this by writing down the samples as follows:

$$(a_1, b_1)\,(a_1, b_2)\,...\,(a_1, b_n)$$
$$(a_2, b_1)\,(a_2, b_2)\,...\,(a_2, b_n)$$
$$...$$
$$(a_m, b_1)\,(a_m, b_2)\,...\,(a_m, b_n).$$

The total number of distinct samples is therefore

$$m \times n. \tag{1}$$

Playing cards are just the special case $m = 4$, $n = 13$.

Another way of stating this result is that if an operation can be performed in m ways and, after it is performed in any one of these ways, a second operation can be performed in n ways, then the two operations can be performed together in

$$m \times n \text{ ways.}$$

This result for two operations (samples of size 2) can be generalized to any number of operations and we state

The multiplication rule. If, for r operations, the ith operation can be performed in n_i ways, the r operations can be performed together in

$$n_1 \times n_2 \times ... \times n_r \text{ ways.} \tag{2}$$

Next, consider samples (x,y) where x and y are chosen, not from two distinct sets S_1 and S_2 as above, but from the *same* set

$$S = \{a_1, a_2, ... a_n\}.$$

Then, from equation (1) we see that there are $n \times n = n^2$ samples of size 2 which can be chosen. What do these samples look like? We list them

$$(a_1, a_1)\,(a_1, a_2)\,...\,(a_1, a_n)$$
$$(a_2, a_1)\,(a_2, a_2)\,...\,(a_2, a_n) \tag{3}$$
$$\cdot \qquad \cdot \quad ... \quad \cdot$$
$$(a_n, a_1)\,(a_n, a_2)\,...\,(a_n, a_n).$$

Two points should be noted.

(i) Repetition of elements is allowed, e.g. (a_2, a_2).

(ii) Sample (a_1, a_2) is counted as being different from sample (a_2, a_1). That is, order of elements in the brackets is important. We call this an *ordered sample*.

Thus, the n^2 samples in (3) correspond to

ordered samples in which repetition is allowed.

Now suppose that repetition of elements is *not* allowed. This means that in (3) we must omit the n diagonal terms. The number remaining is $n^2 - n = n(n - 1)$. We could have arrived at this result by way of the multiplication rule, since there are n choices for x, and once x is chosen, there are $(n - 1)$ choices left for y, giving the total number of samples as $n(n - 1)$. These correspond to

ordered samples in which repetition is not allowed.

The above cases deal with ordered samples, so that the samples (a_1, a_2) and (a_2, a_1) are treated as being different, rather like points in the (x,y)-plane. But some problems which involve selecting two elements are not concerned about the *order* of the elements. For example, a hand of cards does not depend on the order of the cards in the hand. Thus the samples (a_1, a_2) and (a_2, a_1) now amount to only one sample which we shall write as $\{a_1, a_2\}$, using set notation as it is exactly what is required for unordered samples. Such a change, along with no repetitions, means that in (3) we omit the diagonal terms and divide the remaining number of samples by two. This gives $\frac{1}{2}n(n - 1)$ distinct samples corresponding to

unordered samples in which repetition is not allowed.

Our next step is to generalize the above results to samples of size r taken from the set of *distinct* elements

$$S = \{a_1, a_2, \ldots a_n\}.$$

Sample spaces with many elements

Case A: ordered samples; repetition allowed.

Application of the multiplication rule gives the number of samples of size r to be

$$n^r \tag{4}$$

Case B: ordered samples; no repetitions.

Here there are n choices for the first element of the sample, $(n-1)$ for the second, and so on till there are $(n-r+1)$ choices for the rth element. From the multiplication rule we find that the number of samples of size r for this case is

$$n(n-1)(n-2)\ldots(n-r+1) \equiv \frac{n!}{(n-r)!}, \tag{5}$$

where $n!$ is factorial n. Equation (5) is sometimes written as ^nP_r and called the *permutation* of n different objects taken r at a time.

For $r = n$, the number in equation (5) becomes

$$n!, \tag{6}$$

and this is the number of ordered samples of size n that can be chosen from a set of n elements, or, equivalently, the number of different ways in which we can arrange a set of n elements.

Case C: unordered samples; no repetitions.

Consider the set (i.e. unordered sample) of r elements

$$\{a_1, a_2, \ldots a_r\}.$$

From equation (6) we see that there are $r!$ ways of arranging these r elements in order. That is

to 1 set of size r there corresponds $r!$ ordered samples.

Let there be N_r sets of size r. Then

to N_r sets each of size r there corresponds $N_r r!$ ordered samples.

28

But we know from equation (5) that the number of ordered samples of size r, which can be chosen from n elements, is

$$n! / (n - r)!$$

Thus

$$N_r r! = \frac{n!}{(n - r)!},$$

giving

$$N_r = \frac{n!}{r!(n - r)!} \equiv \binom{n}{r}, \qquad (7)$$

where $\binom{n}{r}$ is the *binomial coefficient* which occurs in the expansion

$$(a + b)^n = \sum_{r=0}^{n} \binom{n}{r} a^r b^{n-r}. \qquad (8)$$

The quantity in (7) is called the number of *combinations* of n different objects taken r at a time.

Cases B and C indicate the difference between permutations and combinations:

> In a permutation, order is important;
>
> in a combination, order is not important.

Example

A problem on birthdays. Consider a group of n people, where $n \leqslant 12$. What is the probability that at least two of them have birthdays in the same month? Also, what is the smallest value of n such that the probability that at least two birthdays fall in the same month is greater than one half?

There are 12 possibilities for the month in which a person's birthday falls and hence by (4) there are 12^n possibilities for the birthdays of n people. Thus, the corresponding sample space S contains 12^n points, each of which is an ordered sample

$$(a_1, a_2, ..., a_n),$$

where a_1 represents the birthday of a first person, a_2 represents the birthday of a second person, and so on. Now assume that all of the 12^n possible outcomes are equally likely and associate probability $1/12^n$ with each sample point.

Consider now the event A in S where $A =$ 'no two of the n people have birthdays in the same month'. For this event the birthday of a first person has 12 possible values, that of a second person has 11 possible values, that of a third 10 possible values, ..., and that of the nth person $(12 - n + 1)$ possible values. Hence, by the multiplication rule, the number of possible samples of n birthdays with no two birthdays in the same month is

$$12 \times 11 \times 10 \times \ldots \times (12 - n + 1),$$

and this gives the number of sample points in A. Hence

$$P[A] = \frac{12 \times 11 \times \ldots \times (12 - n + 1)}{12^n}$$

The probability that at least two people have birthdays in the same month is then

$$P[A'] = 1 - P[A] = 1 - \frac{12 \times 11 \times \ldots \times (12 - n + 1)}{12^n}.$$

This number increases as n increases and it is found that for $n = 4$, $P[A'] = 0.43$, and for $n = 5$, $P[A'] = 0.61$. So with a group of only 5 people there is more than an even chance that two of them have birthdays in the same month.

3.3. Subsets and binomial coefficients

The result proved in equation (7) above can be stated as follows:

Theorem 1. The number of subsets of size r that may be formed from the members of a set of size n is $\binom{n}{r}$.

We note here two important properties of the binomial coefficient $\binom{n}{r}$:

$$\binom{n}{r} = \binom{n}{n-r}, \tag{9}$$

$$\binom{n}{r-1} + \binom{n}{r} = \binom{n+1}{r}. \tag{10}$$

These results are readily proved by using the definition of $\binom{n}{r}$ given in equation (7).

Theorem 1 can be used to find the total number of subsets of a set of size n. For there is $\binom{n}{0} \equiv 1$ empty subset, $\binom{n}{1}$ subsets of size 1, $\binom{n}{2}$ of size 2, and so on, up to $\binom{n}{n}$ of size n which is the whole set. So, including the empty set and the whole set, the total number of subsets is

$$\binom{n}{0} + \binom{n}{1} + \binom{n}{2} + \dots + \binom{n}{n}, \qquad (11)$$

and by equation (8), with $a = b = 1$, we see that (11) is equal to

$$2^n. \qquad (12)$$

3.4. Permutations involving identical objects

In equation (6) we saw that the number of different ways in which we can arrange a set of n distinct elements is $n!$. We now consider how this is changed if some of the n elements are identical. For instance, let us find the number of different ways in which we can arrange the letters in the word '*added*'. If the d's were different from one another in this word, the answer would be 5!.

Suppose the unknown total number of permutations of the letters in '*added*' is N. Consider one such permutation; for example,

$$a\,e\,d\,d\,d.$$

If we replace the three d's in this arrangement by

$$d_1, d_2, d_3,$$

the original arrangement gives rise to 3! arrangements by permuting d_1, d_2 and d_3 without disturbing the other letters. In the same way, each of the original N permutations gives rise to 3! permutations. Thus the total number of permutations, with the d's treated as if they are not identical, is

$$N(3!).$$

But the total number of permutations of the five letters

$$a \, e \, d_1 \, d_2 \, d_3$$

is 5!. Thus

$$N(3!) = 5!$$

giving

$$N = \frac{5!}{3!}. \tag{13}$$

This result can be generalized to show that the number of permutations of a set of n elements, where r of the elements are alike and the rest are different, is $\dfrac{n!}{r!}$. Repeated applications of this result give the following theorem.

Theorem 2. The number of different ways in which we can arrange a set of n elements, having n_1 identical elements of one kind, n_2 identical elements of another kind, and so on for k kinds of elements, is

$$\frac{n!}{n_1! \, n_2! \, \ldots \, n_k!}, \tag{14}$$

where $n_1, n_2, \ldots, n_k$ are non-negative integers such that

$$n_1 + n_2 + \ldots + n_k = n. \tag{15}$$

The result in equation (13) is obtained from (14) with $n = 5$, $n_1 = 3$, $n_2 = 1$ and $n_3 = 1$.

EXERCISES FOR CHAPTER THREE

1. How many subsets of size 4 does a set of size 6 possess? How many subsets of all possible sizes does a set of size 6 possess?

2. Find the number of different sets of initials which can be formed if every person has one surname and

        (a) exactly two forenames,
        (b) at most three forenames. (Feller)

3. A certain code contains letters which are formed by a succession of dots and dashes with repetitions allowed. If up to eight successive symbols (dots or dashes) can be used to denote a letter, find the number of different letters that can be formed. (Feller)

4. An instrument maker has 10 types of blanks for constructing a certain lever mechanism. Each blank has 5 different positions where metal can be removed and there are 3 cutting depths at each position except the last, which has only 2. Assuming that cutting at some positions need not occur, but that each blank is cut at least once, find the number of possible levers that can be made.

5. Find the number of different arrangements which can be formed from the letters α, α, β, β, γ.

6. A group of theatres gives both afternoon and evening performances. On a certain day a man wishes to attend two performances, one in the afternoon and one in the evening. There are seven shows that he might consider attending. Find the number of ways in which he can choose two shows. [Hint: consider four cases.]

7. Consider the 24 possible arrangements of the symbols 1, 2, 3, 4, and assign probability $\frac{1}{24}$ to each arrangement. Let A_i be the event that the digit i appears in the ith place from the left in an arrangement (where $i = 1, 2, 3, 4$). Verify that

$$P[A_i \cup A_j] = P[A_i] + P[A_j] - P[A_i \cap A_j]$$

for $i,j = 1, 2, 3, 4$. (Feller)

8. The numbers 1, 2, 3, 4 are arranged in random order. Assuming equally likely outcomes, find the probability that the digits (a) 1 and 2, (b) 1, 2 and 3 appear as neighbours in the order stated.

9. The numbers 1, 2, ..., n are arranged in random order. Assuming equally likely outcomes, find the probabilities of the events (a) and (b) given in exercise 8.

10. A set consists of ten objects numbered 1 to 10. Three objects are selected at random. Find the probability that

(a) the largest,

(b) the smallest

object number chosen will be a 5.

11. From the digits 1, 2, 3, 4 first one is chosen, and then a second selection is made from the remaining three digits. Assume that all possible outcomes have the same probability. Find the probability that an odd digit will be selected (a) the first time, (b) the second time, (c) both times. (Feller)

12. A coin is tossed until for the first time the same result appears twice in succession. Let the probability be $\frac{1}{2^{n-1}}$ for an outcome requiring n tosses. Find the probability that (a) the experiment ends before the fifth toss, (b) an even number of tosses is required. (Feller)

CHAPTER FOUR

The Binomial and Poisson Distributions

4.1. Bernoulli trials

Some experiments consist of repetitions of a particular experiment, that is, they are made up of repeated trials. If the result of each trial is independent of the results of any other trials, we have repeated independent trials.

Definition. Bernoulli trials. Repeated independent trials are called Bernoulli trials if there are only *two* outcomes for each trial and their probabilities remain the same throughout the trials.

The probabilities of the two possible outcomes will be written as p and q, and it is usual to refer to the outcome with probability p as 'success' and to the other as 'failure'. The quantities p and q satisfy

$$p + q = 1, p \geqslant 0, q \geqslant 0. \tag{1}$$

The sample space of each individual trial is $S = \{s,f\}$, s for success and f for failure. The sample space for an experiment consisting of n Bernoulli trials contains 2^n points. Each of these points represents one possible outcome of the n-trial experiment and is made up of n symbols s or f. Since the trials are independent the probabilities multiply (see section 2.6). So, for example, the probability of the outcome $sffs \ldots sfs$ is given by $pqqp \ldots pqp$.

A familiar example of Bernoulli trials is given by the repeated tossing of a coin, for which $p = q = \frac{1}{2}$ if the coin is evenly balanced. If the coin is unbalanced, we can still assume that the successive tosses are independent so that we have a model of Bernoulli trials in which the probability p for success can have any value between 0 and 1.

4.2. The binomial distribution

Frequently one is interested in the total number of successes obtained in a succession of n Bernoulli trials but not in their order. The number of successes can be $0, 1, ..., n$, and the problem is to determine the corresponding probabilities. Suppose there are k successes and $n - k$ failures in a sequence of n Bernoulli trials. If the order in which the successes occur does not matter, there are $\binom{n}{k}$ sample points belonging to this event, and each of these sample points occurs with probability $p^k q^{n-k}$. Thus by axiom III of chapter 2, the probability of k successes in n trials is

$$b(k; n,p) = \binom{n}{k} p^k q^{n-k}, k = 0, 1, ..., n. \qquad (2)$$

The set of probabilities $b(k; n,p)$ is called the *binomial distribution*.

Adding all the probabilities $b(k; n,p)$ we get

$$\sum_{k=0}^{n} b(k; n,p) = \sum_{k=0}^{n} \binom{n}{k} p^k q^{n-k}$$

$$= (p + q)^n \text{ by the binomial theorem,}$$

$$= 1,$$

since $p + q = 1$. This result shows that we have accounted for all the probabilities in the sample space.

Example

A die is thrown nine times. Let k be the number of throws for which the outcome is a six. Find the value of k for which $b(k; n, p)$ is a maximum.

Let $$p = \text{probability of a six}$$
and $$q = 1 - p.$$

Then, the probability of getting a six k times in nine throws is

$$b(k; 9,p) = \binom{9}{k} p^k q^{9-k}.$$

Assuming that $p = \dfrac{1}{6}$, we have that $q = \dfrac{5}{6}$, and so

$$\frac{b\left(k + 1; 9, \frac{1}{6}\right)}{b\left(k; 9, \frac{1}{6}\right)} = \frac{1}{5}\frac{9 - k}{1 + k},$$

which gives

$$b\left(1; 9, \frac{1}{6}\right) > b\left(0; 9, \frac{1}{6}\right),$$

and

$$b\left(1; 9, \frac{1}{6}\right) > b\left(2; 9, \frac{1}{6}\right) > \ldots > b\left(9; 9, \frac{1}{6}\right).$$

Thus

$$b\left(k; 9, \frac{1}{6}\right) \text{ is a maximum for } k = 1.$$

4.3. The central term

The example we have just considered can be generalized as follows. From (2) we see that

$$\frac{b(k; n,p)}{b(k - 1; n,p)} = \frac{(n - k + 1)p}{kq} = 1 + \frac{(n + 1)p - k}{kq}, \quad (3)$$

using $p = 1 - q$. Hence, the term $b(k; n,p)$ satisfies

$$b(k; n,p) > b(k - 1; n,p) \quad \text{if } (n + 1)p > k,$$

and $$b(k; n,p) < b(k - 1; n,p) \quad \text{if } (n + 1)p < k.$$

If $(n + 1)p$ happens to be an integer m, then for $k = m$, (3)
gives $$b(m; n,p) = b(m - 1; n,p).$$

Now there is exactly one integer m such that

$$(n + 1)p - 1 < m \leqslant (n + 1)p. \qquad (4)$$

We can therefore state

Theorem 1. As k goes from 0 to n, the terms $b(k; n,p)$ first increase monotonically, then decrease monotonically, reaching their maximum when $k = m \equiv [(n + 1)p]$, where $[x]$ denotes the largest integer not exceeding x. When $(n + 1)p$ is an integer, $m = (n + 1)p$, and in this case $b(m - 1; n,p) = b(m; n,p)$.

The term $b(m; n,p)$, where $m = [(n + 1)p]$, is called the *central* term.

Consider equation (3) again. If r is any positive integer, and $k \geqslant r + 1$, then

$$\frac{(n - k + 1)p}{kq} \leqslant \frac{(n - r)p}{(r + 1)q}$$

so that

$$\frac{b(k; n,p)}{b(k - 1; n,p)} \leqslant \frac{(n - r)p}{(r + 1)q} = \beta, (k \geqslant r + 1). \qquad (5)$$

From (5) we then get for any positive integer i

$$\frac{b(r + i; n,p)}{b(r + i - 1; n,p)} \cdot \frac{b(r + i - 1; n,p)}{b(r + i - 2; n,p)} \cdots \frac{b(r + 1; n,p)}{b(r; n,p)} < \beta^i$$

that is,

$$\frac{b(r + i; n,p)}{b(r; n,p)} \leqslant \beta^i, (i = 1, ..., n - r). \qquad (6)$$

For $r \geqslant np$, $\beta = \dfrac{(n - r)p}{(r + 1)q} < 1$. Then from (6)

$$\sum_{i=0}^{n-r} b(r + i; n,p) \leqslant b(r; n,p) \sum_{i=0}^{n-r} \beta^i$$

$$= b(r; n,p) \frac{1 - \beta^{n-r+1}}{1 - \beta}, \beta < 1,$$

$$\leqslant b(r; n,p) \; \frac{1}{1 - \beta}$$

$$= b(r; n,p) \; \frac{(r + 1)q}{(r + 1)q - (n - r)p},$$

and using $q = 1 - p$,

$$= b(r; n,p) \; \frac{(r + 1)q}{r + 1 - (n + 1)p}.$$

Thus, for $r \geqslant np$, we have the relation

$$\sum_{i=0}^{n-r} b(r + i; n,p) \leqslant b(r; n,p) \frac{(r + 1)q}{r + 1 - (n + 1)p}. \tag{7}$$

Equation (7) accounts for the probabilities $b(r; n,p)$, $b(r + 1; n,p)$, ..., up to $b(n; n,p)$.

We can readily find a similar relation which accounts for the probabilities $b(k; n,p)$ for $k = 0, 1, ..., r - 1$. If we interchange p and q in equation (7) we have

$$\sum_{i=0}^{n-r} b(r + i; n,q) \leqslant b(r; n,q) \frac{(r + 1)p}{r + 1 - (n + 1)q} (r \geqslant nq). \tag{8}$$

Now it follows from the definition (2) of $b(k; n,p)$ that

$$b(k; n,p) = b(n - k; n,q). \tag{9}$$

Using this in (8) we have

$$\sum_{i=0}^{n-r} b(n - r - i; n,p) \leqslant b(n - r; n,p)$$

$$\frac{(r + 1)p}{r + 1 - (n + 1)q} (r \geqslant nq).$$

If we set $s = n - r$, then $s \leqslant n - nq = np$, and we have

$$\sum_{i=0}^{s} b(s - i; n,p) \leqslant b(s; n,p)$$

$$\frac{(n - s + 1)p}{n - s + 1 - (n + 1)q} (s \leqslant np)$$

or

$$\sum_{i=0}^{s} b(i;\, n,p) \leqslant b(s;\, n,p)\, \frac{(n-s+1)p}{(n+1)p-s}\, (s \leqslant np), \qquad (10)$$

where on the left hand side of (10) the terms simply appear in the opposite order to those in the previous line.

We can state results (7) and (10) as follows.

Theorem 2. If $r \geqslant np$, the probability of at least r successes (in a sequence of n Bernoulli trials) satisfies the inequality (7); if $s \leqslant np$, the probability of at most s successes satisfies the inequality (10).

4.4. The law of large numbers

If a coin is tossed once and we choose the probability model

$$S = \{H,T\},$$
$$P[\{H\}] = P[\{T\}] = \tfrac{1}{2},$$

then the intuitive picture which we have is that if the coin is tossed a great many times, 'heads' should be expected, in the long run, to occur fifty times out of a hundred. This is what is known as the relative frequency interpretation of probability.

For any sequence of Bernoulli trials we can state the intuitive notion of probability as follows. If the number of successes in the first n trials in a sequence of Bernoulli trials is S_n, then

$$\frac{S_n}{n} \text{ should be near } p, \qquad (11)$$

where p is the probability of success in one trial. In the abstract theory this cannot be true for *every* sequence of trials, since in fact the sample space contains a point which corresponds to an uninterrupted sequence of successes, and for it $S_n/n = 1$. Nevertheless, there is some sense in which (11) is true.

Consider the probability that S_n/n exceeds $p + \epsilon$, where $\epsilon > 0$ is arbitrarily small but fixed. This probability is the same as $P[S_n > n(p + \epsilon)]$ and is equal to the left hand side of (7) where r is the smallest integer exceeding $n(p + \epsilon)$. Then (7) implies

$$P[S_n > n(p + \epsilon)] < b(r; n,p) \frac{n(p + \epsilon)q + q.}{n\epsilon + q} \qquad (12)$$

As n increases, the fraction on the right remains bounded, whereas $b(r; n,p)$ tends to zero*. Thus

$$P[S_n > n(p + \epsilon)] \to 0 \text{ as } n \to \infty.$$

Using (10) in the same way we see that

$$P[S_n < n(p - \epsilon)] \to 0 \text{ as } n \to \infty.$$

These two results may be stated as

$$P\left[|\frac{S_n}{n} - p| > \epsilon \right] \to 0 \text{ as } n \to \infty. \qquad (13)$$

So, as n increases, the probability that the average number of successes deviates from p by more than any preassigned ϵ tends to zero. This is one form of the *law of large numbers* and provides a basis for the relative frequency interpretation of probability.

4.5. The Poisson approximation

In many cases of Bernoulli trials, n is large and p is comparatively small, whereas the product

$$\lambda = np \qquad (14)$$

is of moderate size. In these cases we can find an approximation to $b(k; n,p)$. This approximation formula is due to Poisson. We have from equation (2)

---

* To prove this use Stirling's formula given on page 59.

$$b(k; n,p) = \binom{n}{k} p^k q^{n-k}$$

$$= \binom{n}{k} \left(\frac{\lambda}{n}\right)^k \left(1 - \frac{\lambda}{n}\right)^{n-k}$$

$$= \frac{1}{k!} \lambda^k \left(1 - \frac{\lambda}{n}\right)^{n-k} \frac{n(n-1) \dots (n-k+1)}{n^k}. \tag{15}$$

So far this is exact. We now introduce an approximation.
For large n,

$$\frac{n(n-1) \dots (n-k+1)}{n^k} = 1 + 0\left(\frac{1}{n}\right) \simeq 1, \tag{16}$$

where the sign $\simeq$ indicates approximate equality, the terms of order of magnitude $\frac{1}{n}$ being neglected in comparison with terms of order of magnitude unity.

Also, consider

$$\log \left(1 - \frac{\lambda}{n}\right)^{n-k} = (n-k) \log \left(1 - \frac{\lambda}{n}\right),$$

$$= (n-k)\left(-\frac{\lambda}{n} - \frac{1}{2}\frac{\lambda^2}{n^2} - \dots\right)$$

by the Taylor expansion,

$$= -\lambda + \frac{1}{n}\left(k\lambda - \frac{1}{2}\lambda^2\right) - \dots.$$

Thus, for large n,

$$\left(1 - \frac{\lambda}{n}\right)^{n-k} \simeq e^{-\lambda}. \tag{17}$$

Using the approximations (16) and (17) in (15) we obtain

$$b(k; n,p) \simeq \frac{\lambda^k}{k!} e^{-\lambda}. \tag{18}$$

This is the *Poisson approximation* to the binomial distribution.

If
$$p(k; \lambda) = \frac{\lambda^k}{k!} e^{-\lambda}, \; k = 0, 1, 2, \ldots, \tag{19}$$

then $p(k; \lambda)$ should be an approximation to $b\left(k; n, \frac{\lambda}{n}\right)$ when n is sufficiently large.

A comparison of $b\left(k; n, \frac{\lambda}{n}\right)$ and $p(k; \lambda)$ is given in the table for $n = 100$ and $\lambda = 1$.

TABLE

k	$b\left(k; 100, \frac{1}{100}\right)$	$p(k; 1)$
0	0·3660	0·3679
1	0·3697	0·3679
2	0·1849	0·1839
3	0·0610	0·0613
4	0·0149	0·0153

These results indicate that the Poisson approximation is satisfactory when n is large and λ is of the order of unity.

4.6. The Poisson distribution

In the preceding section we discussed the Poisson expression (19) and saw that it is a useful approximation to the binomial distribution (2) when n is large and p is small.

However, since for a fixed value of λ

$$\sum_{k=0}^{\infty} p(k; \lambda) = \sum_{k=0}^{\infty} \frac{\lambda^k}{k!} e^{-\lambda} = 1, \tag{20}$$

we can regard $p(k; \lambda)$ as the probability of exactly k successes in some, perhaps ideal, experiment. That is,

$$p(k; \lambda) = \frac{\lambda^k}{k!} e^{-\lambda}, \; k = 0, 1, 2, \ldots, \tag{21}$$

is a probability distribution in its own right. It is called the *Poisson distribution*. It turns out that there are many physical experiments which lead to just this distribution. For example, (21) in various forms describes radioactive disintegrations, and queuing problems such as incoming calls at a telephone exchange.

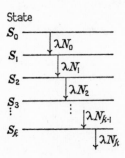

State

A decay process

We now consider how the Poisson distribution arises in a simple radioactive decay process. Suppose we have N atoms or particles and at time $t = 0$ they all have a property labelled by S_0. We say that they are all in *state* S_0. As time goes on some of the particles may lose this property (i.e. leave the state S_0) by emitting energy, and drop down to state S_1. Similarly, some of the particles which reach state S_1 may lose energy and drop down to state S_2, and so on. This is a simple decay process. A mathematical model for it can be found by supposing that the rate of loss of particles from a given state S_k is proportional to the number of particles N_k in the state. For state S_0 this means that

$$\frac{\mathrm{d}N_0}{\mathrm{d}t} = -\lambda N_0, \qquad (22)$$

where λ is some positive constant. In (22) we are treating N_0 as a differentiable function of t to simplify the discussion. For state S_k we might suppose similarly that

$$\frac{\mathrm{d}N_k}{\mathrm{d}t} = - \lambda N_k.$$

However, in this expression we have forgotten that particles are arriving in state S_k from state S_{k-1} at the rate of λN_{k-1}. So this equation must be modified to read

$$\frac{\mathrm{d}N_k}{\mathrm{d}t} = - \lambda N_k + \lambda N_{k-1}, (k = 1, 2, ...). \qquad (23)$$

If $N_0 = N$, $N_1 = N_2 = ... = 0$ at $t = 0$, these equations (22) and (23) are satisfied by*

$$N_k = N \frac{(\lambda t)^k}{k!} \mathrm{e}^{-\lambda t}, (k = 0, 1, 2, ...,). \qquad (24)$$

Equation (24) gives the number of particles in state S_k at time t. If N is very large, the work of section 4.4 suggests that in a certain sense, the ratio N_k/N should be near the probability $p(k; \lambda t)$ that N_k particles reach S_k at time t. Thus

$$p(k; \lambda t) \simeq \frac{N_k}{N} = \frac{(\lambda t)^k}{k!} \mathrm{e}^{-\lambda t}, \qquad (25)$$

which is just the Poisson distribution (21), with λt instead of λ.

4.7. Generating functions

We mention briefly here the notion of generating functions which play an important part in more advanced work in probability theory.

Definition. Let $p_0, p_1, p_2, ...$ be a sequence of real numbers. If

$$A(s) = p_0 + p_1 s + p_2 s^2 + ... \qquad (26)$$

converges in some interval $|s| < s_0$, then $A(s)$ is called the generating function of the sequence $\{p_j\}$.

---

* For a treatment of this see, for example, G. E. H. Reuter: *Elementary Differential Equations and Operators*, Routledge & Kegan Paul, Library of Mathematics.

In (26) the variable s itself has no significance, but the *power* of s and the coefficients p_j are important. For our purposes the numbers p_j represent probabilities.

For example, consider the throw of a die and let j be the number scored. Then the only non-zero p_j are

$$p_j = \frac{1}{6}, j = 1, 2, ..., 6,$$

assuming the sample points are equally likely. Putting these values in (26) we have

$$A(s) = \frac{1}{6}(s + s^2 + s^3 + s^4 + s^5 + s^6) \qquad (27)$$

as the corresponding generating function.

A second example is provided by the binomial distribution with probabilities

$$b(k; n, p) = \binom{n}{k} p^k q^{n-k}, k = 0, 1, ..., n. \qquad (28)$$

We find the generating function of the sequence $\{b(k; n, p)\}$ by writing

$$\begin{aligned}
A(s) &= \sum_{k=0}^{n} b(k; n, p) s^k \\
&= \sum_{k=0}^{n} \binom{n}{k} (ps)^k q^{n-k} \\
&= (q + ps)^n.
\end{aligned} \qquad (29)$$

Thus, the probability of having k successes in n Bernoulli trials is given by the coefficient of s^k in (29).

As another example consider the Poisson distribution

$$p(k; \lambda) = \frac{\lambda^k e^{-\lambda}}{k!}, k = 0, 1, 2, ... \qquad (30)$$

The generating function of the sequence $\{p(k; \lambda)\}$ is

$$
\begin{aligned}
A(s) &= \sum_{k=0}^{\infty} p(k; \lambda)s^k \\
&= \sum_{k=0}^{\infty} \frac{(\lambda s)^k e^{-\lambda}}{k!} \\
&= e^{-\lambda + \lambda s}.
\end{aligned}
\tag{31}
$$

If we are given a generating function $P(s) = \sum_k p_k s^k$, then the probabilities p_k can be obtained from the formula

$$
p_k = \frac{P^{(k)}(0)}{k!},
\tag{32}
$$

where $\qquad P^{(k)}(0) = \dfrac{\mathrm{d}^k P}{\mathrm{d}s^k}$ at $s = 0$.

EXERCISES FOR CHAPTER FOUR

1. If an ordinary die is thrown four times, what is the probability that exactly two sixes will occur?

2. If the probability of hitting a target is 0.4 and five shots are fired, what is the probability that the target will be hit at least twice?

3. In exercise 2, find the conditional probability that the target will be hit at least twice, assuming that at least one hit is scored.

4. If a bullet has probability p of hitting its target, find the probability that in n shots there will be (i) no hits, (ii) exactly one hit, (iii) at least two hits.

5. A fair coin tossed six times constitutes one trial of an experiment. In a sequence of such trials what proportion of the outcomes will contain three heads and three tails?

6. One trial of an experiment consists in throwing two fair dice independently and the possible outcomes are represented by (i, j) with $i, j = 1, 2, \ldots, 6$. If a sequence of n such trials is carried out, find the probability that the nth trial will be the first one for which $i + j = 9$.

7. A true coin is tossed $2n$ times. Find the probability that the first n tosses and the second n tosses result in the same number of heads.

8. A sequence of Bernoulli trials is performed until the first success occurs. For any integer $n = 1, 2, \ldots$, find the probability that n will be the number of trials required to achieve the first success. (Parzen)

9. Suppose that there are k successes in n Bernoulli trials. Show that the conditional probability that any particular trial resulted in a success is equal to k/n. (Parzen)

10. Prove that the probability $b(r; n, p)$ in equation (12) tends to zero as n tends to infinity.

11. As k takes the values $0, 1, 2, \ldots$, show that the probabilities of the Poisson distribution

$$p(k; \lambda) = \frac{\lambda^k e^{-\lambda}}{k!}$$

first increase monotonically, and then decrease monotonically, reaching their greatest value when k is the largest integer not exceeding λ.

12. The probability that a given box contains k red balls is given by the Poisson distribution $p(k; \lambda)$ and the probability that the box contains r white balls is $p(r; \mu)$, the two events being assumed independent. Show that the probability that the box contains a total of n balls is $p(n; \lambda + \mu)$. [Hint: the n balls may consist of k red ones and $r = n - k$ white ones, where k takes the values $0, 1, \ldots, n$.]

CHAPTER FIVE

Probability and Continuous Sample Spaces

5.1. Introduction

So far we have dealt with sample points that could be counted off in terms of the positive integers. That is, the sample spaces have been discrete. However, there are many sample spaces with sample points lying in an interval or even over the whole real line. An example of this is provided by the sample space corresponding to the position of a particle moving along a straight line.

If a sample space is the real line or intervals on the real line, we call it *continuous*. Of course, it is possible to think of extending a discrete sample space, such as $S = \{1,2,3,4,5,6\}$, to be the whole of the real line. The probability zero would then be attached to all points except at the integers 1 to 6. However, we are concerned here with sample spaces which describe continuous variables, rather than with the extension of sample spaces that are basically discrete.

5.2. Continuous probability distributions

We now consider how probability is defined on continuous sample spaces. For sample spaces such as the entire real line $(-\infty, \infty)$, we introduce the function

$$F(x) = P[\{x'; x' \leqslant x\}], \tag{1}$$

which gives the probability that the sample point has any value less than or equal to a specified value x. $F(x)$ is called the *distribution function*.

48

If we take x large enough we will in the limit have the certain event S, so that

$$\lim_{x \to +\infty} F(x) = F(+\infty) = 1. \tag{2}$$

If we let x tend to $-\infty$ we will in the limit have the null event, so that

$$\lim_{x \to -\infty} F(x) = F(-\infty) = 0. \tag{3}$$

Also, let $a < b$ and consider

$$\begin{aligned} F(b) - F(a) &= P[\{x'; x' \leqslant b\}] - P[\{x'; x' \leqslant a\}] \\ &= P[\{x'; a < x' \leqslant b\}], \end{aligned} \tag{4}$$

and this expression, being a probability, is non-negative. Thus $F(x)$ is a monotonic non-decreasing function of x.

We now restrict ourselves to distribution functions $F(x)$ that are *continuous* and have a *derivative at all points* x of the real line. In such cases we call the probability distribution *continuous*.

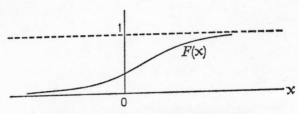

Fig. 5.1. A continuous distribution function

Let the derivative of $F(x)$ be

$$\frac{\mathrm{d}}{\mathrm{d}x} F(x) = f(x). \tag{5}$$

Since $F(x)$ is a non-decreasing function, we have

$$f(x) \geqslant 0. \tag{6}$$

E

Then from the fundamental theorem of integral calculus, we obtain

$$F(x) - F(-\infty) = \int_{-\infty}^{x} f(x')\mathrm{d}x',$$

and using (3), this becomes

$$F(x) = \int_{-\infty}^{x} f(x')\mathrm{d}x'. \tag{7}$$

We use x' as the variable of integration to avoid confusion with x which is the upper limit in (7). The function $f(x)$ is called a *probability density function*, and we shall now consider its properties.

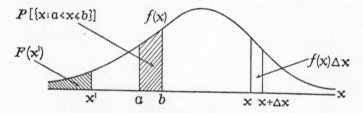

FIG. 5.2. A probability density function

5.3. Probability density functions

An example of a probability density function $f(x)$ is shown in figure 5.2. We have already seen in (6) that $f(x)$ is non-negative. The probability of the event $\{x'; a < x' \leqslant b\}$ is given by (4) as

$$P[\{x'; a < x' \leqslant b\}] = F(b) - F(a)$$

$$= \int_{-\infty}^{b} f(x)\mathrm{d}x - \int_{-\infty}^{a} f(x)\mathrm{d}x, \quad \text{by (7)},$$

$$= \int_a^b f(x)\mathrm{d}x, \qquad (8)$$

which is the area under the curve $f(x)$ between $x = a$ and $x = b$. Also, the probability of the event

$$\{x'; x < x' \leqslant x + \varDelta x\}$$

is given by

$$\int_x^{x+\varDelta x} f(x')\mathrm{d}x',$$

which is approximately

$$f(x)\varDelta x \qquad (9)$$

when $\varDelta x$ is small. Thus, to find the probability of x lying in the range $(x, x+\varDelta x)$ we take the area under the curve between x and $x + \varDelta x$ and obtain approximately $f(x)\varDelta x$. Hence, $f(x)$ by itself *does not* represent probability, but $f(x)$ times the length of a short interval *does* give probability. For this reason $f(x)$ is called a probability *density*.

From (2) and (7) we have

$$\int_{-\infty}^{\infty} f(x)\mathrm{d}x = 1. \qquad (10)$$

Equations (6) and (10) are the two conditions for $f(x)$ to be a density function. It should be noted that it is possible for a density function itself to take values greater than 1.

Now we consider the question: What is the probability of getting a value x exactly? From above we have seen that the probability of x lying in the range $(x, x + \varDelta x)$ is $f(x)\varDelta x$. As $\varDelta x$ decreases, so that the interval $(x, x + \varDelta x)$ reduces to the point x, the probability $f(x)\varDelta x$ tends to zero. Thus, the probability of getting precisely the value x is zero. This may seem a strange result at first. The reason is that in any experiment we can only record measurements to perhaps three or four decimal places, so that the corresponding sample space is really discrete. Our treatment by means of continuous variables is an idealization of this sample space in which the result x is a real number with its full decimal expansion. Hence we would expect the probability zero for obtaining a value x exactly.

Example 1 Consider the function

$$f(x) = 2x - x^2 \qquad 0 \leqslant x \leqslant 2,$$
$$= 0 \qquad \text{elsewhere.} \tag{11}$$

This function is non-negative. The integral

$$\int_{-\infty}^{\infty} f(x)\mathrm{d}x = \int_{0}^{2} (2x - x^2)\mathrm{d}x = \frac{4}{3} \neq 1.$$

Thus, $f(x)$ in (11) is *not* a probability density function as it stands, but the function

$$f(x) = \frac{3}{4}(2x - x^2) \qquad 0 \leqslant x \leqslant 2,$$
$$= 0 \qquad \text{elsewhere,} \tag{12}$$

is a probability density function.

Example 2 To compute probabilities from a density function.

Consider the function

$$\begin{aligned} f(x) &= 0 & x &\leqslant 0 \\ &= \frac{1}{5}(x + 1) & 0 &< x \leqslant 1 \\ &= \frac{2}{5} & 1 &\leqslant x \leqslant 2 \\ &= \frac{1}{5}(4 - x) & 2 &\leqslant x \leqslant 3 \\ &= 0 & x &> 3. \end{aligned} \tag{13}$$

This function satisfies conditions (6) and (10), and so is a probability density function.

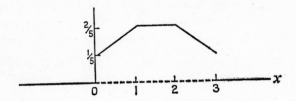

Fig. 5.3. $f(x)$ of equation (13)

We now define two events A and B.

$$A = \{x; 0 < x \leqslant 2\} \text{ and } B = \{x; 1 < x \leqslant 3\}.$$

Then

$$P[A] = \int_0^2 f(x)\mathrm{d}x = \int_0^1 \frac{1}{5}\,(x+1)\mathrm{d}x + \int_1^2 \frac{2}{5}\,\mathrm{d}x = \frac{7}{10}$$

$$P[B] = \int_1^3 f(x)\mathrm{d}x = \int_1^2 \frac{2}{5}\mathrm{d}x + \int_2^3 \frac{1}{5}(4-x)\mathrm{d}x = \frac{7}{10}.$$

Also, $A \cap B = \{x; 1 < x \leqslant 2\}$, and so

$$P[A \cap B] = \int_1^2 f(x)\mathrm{d}x = \int_1^2 \frac{2}{5}\,\mathrm{d}x = \frac{2}{5}.$$

5.4. The uniform distribution

Consider the finite interval (a, b) of the sample space $S = (-\infty, \infty)$ and introduce events A and B as follows:

$$A = \{x'; a \leqslant x' \leqslant b\},$$
$$\text{and } B = \{x'; a \leqslant x' \leqslant x\}.$$

Then we define the uniform probability distribution by taking

53

$$P[B] = \frac{\text{length of } B}{\text{length of } A} \quad \text{if } B \subset A,$$

$$= 0 \quad \text{if } B \not\subset A, \tag{14}$$

where by 'length' we mean the length of the interval of the real line corresponding to the event. The distribution function for this is

$$F(x) = 0 \qquad x < a,$$

$$= \frac{x-a}{b-a} \qquad a \leqslant x \leqslant b,$$

$$= 1 \qquad x > b, \tag{15}$$

and the density function is, from (5) and (15),

$$f(x) = \frac{1}{b-a} \qquad a \leqslant x \leqslant b,$$

$$= 0 \qquad \text{otherwise.} \tag{16}$$

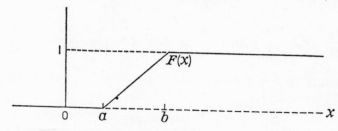

FIG. 5.4. The uniform distribution function (15)

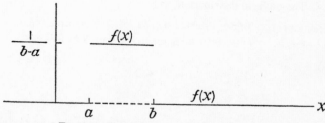

FIG. 5.5. The uniform density function (16)

54

From (14) it can be seen that the notion of the uniform distribution represents an extension of the notion of a finite sample space S with equally likely outcomes (see section 2.4).

5.5. The normal distribution

A continuous distribution of great importance in probability theory is the so-called normal distribution. This distribution involves two parameters μ and σ, where σ must be positive. The density function is

$$f(x) = \frac{1}{\sigma\sqrt{2\pi}} \exp\left[-\tfrac{1}{2}\left(\frac{x-\mu}{\sigma}\right)^2\right], \tag{17}$$

and the distribution function is then

$$F(x) = \frac{1}{\sigma\sqrt{2\pi}} \int_{-\infty}^{x} \exp\left[-\tfrac{1}{2}\left(\frac{y-\mu}{\sigma}\right)^2\right]dy, \tag{18}$$

where we use y as the variable of integration to avoid confusion with x, which is the upper limit in (18). The graph of the normal density function is the symmetric, bell-shaped curve shown in figure 5.6. Figure 5.7 shows the corresponding distribution function.

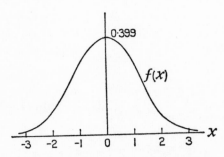

FIG. 5.6. The normal density function
with $\mu = 0$, $\sigma = 1$

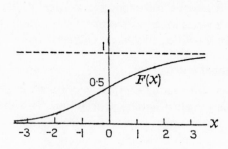

Fig. 5.7. The normal distribution function with $\mu = 0$, $\sigma = 1$

We now check that $f(x)$ is, in fact, a density function. First, $f(x)$ is non-negative, since we assumed σ positive. Second, we must show that the area under $f(x)$ is 1. The area is

$$A = \int_{-\infty}^{\infty} f(x)\mathrm{d}x = \frac{1}{\sigma\sqrt{2\pi}} \int_{-\infty}^{\infty} \exp\left[-\tfrac{1}{2}\left(\frac{y-\mu}{\sigma}\right)^2\right]\mathrm{d}y$$

$$= \frac{1}{\sqrt{2\pi}} \int_{-\infty}^{\infty} \exp\left(-\tfrac{1}{2}z^2\right)\mathrm{d}z, \qquad (19)$$

where we have put $z = (y-\mu)/\sigma$. This integral A is clearly positive. We prove that it is equal to 1 by showing that its square is equal to 1.

We have

$$A^2 = \frac{1}{2\pi} \int_{-\infty}^{\infty} \exp\left(-\tfrac{1}{2}z^2\right)\mathrm{d}z \int_{-\infty}^{\infty} \exp\left(-\tfrac{1}{2}v^2\right)\mathrm{d}v$$

$$= \frac{1}{2\pi} \int_{-\infty}^{\infty} \int_{-\infty}^{\infty} \exp\left(-\frac{z^2+v^2}{2}\right)\mathrm{d}z\,\mathrm{d}v. \qquad (20)$$

We evaluate this by writing

$$z = r \cos \theta,$$
$$v = r \sin \theta.$$

Then the differential element of area $dz\, dv$ becomes $r\, dr\, d\theta$, and

$$A^2 = \frac{1}{2\pi} \int_0^{2\pi} \int_0^\infty \exp\left(-\tfrac{1}{2}r^2\right) r\, dr\, d\theta$$

$$= \int_0^\infty r \exp\left(-\tfrac{1}{2}r^2\right) dr, \text{ and setting } s = \tfrac{1}{2}r^2$$

$$= \int_0^\infty \exp\left(-s\right) ds$$

$$= 1, \tag{21}$$

which is the required result.

When $\mu = 0$ and $\sigma = 1$, the distribution is called the *standardized* normal distribution with density function

$$f_s(x) = \frac{1}{\sqrt{2\pi}} \exp\left(-\tfrac{1}{2}x^2\right), \tag{22}$$

and distribution function

$$F_s(x) = \frac{1}{\sqrt{2\pi}} \int_{-\infty}^x \exp\left(-\tfrac{1}{2}y^2\right) dy. \tag{23}$$

It is readily shown that these functions have the properties

$$f_s(-x) = f_s(x), \tag{24}$$
$$F_s(-x) = 1 - F_s(x). \tag{25}$$

So, once the functions are tabulated for positive values of x, equations (24) and (25) give their values for negative x.

The importance of the normal distribution in probability theory arises because in a certain limit it approximates the binomial distribution discussed in Chapter 4. We now consider this approximation.

5.6. The normal approximation to the binomial distribution

Let S_n be the number of successes in n Bernoulli trials with probability p of success in each trial. Then (equation (2), Chapter 4)

$$b(k; n,p) = \binom{n}{k} p^k q^{n-k} \tag{26}$$

is the probability of the event $S_n = k$. We now find an approximation to $b(k; n,p)$ in the limit when $n \to \infty$, but p is kept fixed. From the law of large numbers, equation (13) of Chapter 4, the probability that

$$|S_n - np| > n\epsilon,$$

tends to zero for each $\epsilon > 0$. So only the values of $S_n = k$ for which

$$|k - np| < n\epsilon,$$

that is,

$$\frac{1}{n} |k - np| \to 0, \tag{27}$$

need be considered. If we write $\xi_k = k - np$, then

$$k = np + \xi_k, \, n - k = nq - \xi_k, \tag{28}$$

and we are interested only in combinations n, k such that $n \to \infty$ and $\xi_k/n \to 0$.

From (26)

$$b(k; n,p) = \frac{n!}{k!(n-k)!} p^k q^{n-k}$$

and using Stirling's formula* for the asymptotic value of a factorial

$$n! \sim (2\pi)^{\frac{1}{2}} n^{n+\frac{1}{2}} e^{-n},$$

where $\sim$ means that the ratio of the two sides tends to unity as $n \to \infty$, we have

$$b(k;n,p) \sim \left\{\frac{n}{2\pi k\ (n-k)}\right\}^{\frac{1}{2}} \frac{n^n}{k^k(n-k)^{n-k}}\ p^k q^{n-k}$$

$$= \left\{\frac{n}{2\pi k\ (n-k)}\right\}^{\frac{1}{2}} \left(\frac{np}{k}\right)^k \left(\frac{nq}{n-k}\right)^{n-k}$$

$$= \left\{\frac{n}{2\pi(np + \xi_k)\ (nq - \xi_k)}\right\}^{\frac{1}{2}}$$

$$\frac{1}{\left(1 + \dfrac{\xi_k}{np}\right)^{np+\xi_k} \left(1 - \dfrac{\xi_k}{nq}\right)^{nq-\xi_k}}. \qquad (29)$$

Let

$$A = \left(1 + \frac{\xi_k}{np}\right)^{np+\xi_k} \left(1 - \frac{\xi_k}{nq}\right)^{nq-\xi_k},$$

and consider

$$\log A = (np + \xi_k) \log \left(1 + \frac{\xi_k}{np}\right) +$$

$$+ (nq - \xi_k) \log \left(1 - \frac{\xi_k}{nq}\right).$$

---

* For details see, for example, W. Ledermann, *Integral Calculus*, Routledge & Kegan Paul, Library of Mathematics.

If $|\xi_k| < npq$ we may expand the logarithms, and so

$$\log A = (np + \xi_k)\left(\frac{\xi_k}{np} - \frac{\xi_k^2}{2n^2p^2} + \frac{\xi_k^3}{3n^3p^3} - \cdots\right) +$$

$$+ (nq - \xi_k)\left(-\frac{\xi_k}{nq} - \frac{\xi_k^2}{2n^2q^2} - \frac{\xi_k^3}{3n^3q^3} - \cdots\right)$$

$$= \frac{\xi_k^2}{2npq}\left\{1 + \frac{p-q}{3pq}\frac{\xi_k}{n} + \cdots\right\}. \tag{30}$$

If we suppose that $\xi_k^3/n^2 \to 0$ then all the terms of (30) except the first tend to zero and so

$$A \sim e^{\xi_k^2/2npq}.$$

Putting this in (29) gives

$$b(k; n,p) \sim \left\{\frac{n}{2\pi(np + \xi_k)(nq - \xi_k)}\right\}^{\frac{1}{2}} e^{-\xi_k^2/2npq}. \tag{31}$$

Now $np + \xi_k \sim np$ and $nq - \xi_k \sim nq$ and so (31) becomes

$$b(k; n,p) \sim \frac{1}{(2\pi npq)^{\frac{1}{2}}} e^{-\xi_k^2/2npq}. \tag{32}$$

Comparison with (17) shows that for large n and $\xi_k^3/n^2 \to 0$ the binomial distribution $b(k; n,p)$ is approximated by the normal distribution.

EXERCISES FOR CHAPTER FIVE

1. Verify that each of the following functions is a probability density function and sketch its graph.

$$\text{(i) } f(x) = \frac{1}{2|\sqrt{x}|} \qquad 0 < x < 1,$$

$$= 0 \qquad \text{elsewhere.}$$

$$(ii)\ f(x) = 2x \qquad 0 < x < 1,$$
$$\quad = 0 \qquad \text{elsewhere.}$$

$$(iii)\ f(x) = |x| \qquad |x| \leqslant 1,$$
$$\quad = 0 \qquad \text{elsewhere.}$$

2. Consider the function

$$f(x) = 6x(1 - x) \qquad 0 < x < 1,$$
$$\quad = 0 \qquad \text{elsewhere.}$$

Check that $f(x)$ is a density function. Evaluate $P[\{x; \frac{2}{3} < x < 1\}]$.

3. Calculate the distribution function $F(x)$ for $f(x)$ in exercise 2.

4. The Cauchy distribution has density function

$$f(x) = \frac{c}{1 + x^2} \qquad -\infty < x < \infty.$$

Evaluate c. Find the distribution function. Calculate

$$P[\{x; 0 < x < 1\}].$$

5. The Pareto distribution has distribution function

$$F(x) = 0 \qquad\qquad x < x_0,$$
$$\quad = 1 - cx^{-\alpha} \qquad x \geqslant x_0,$$

where α is a positive parameter and x_0 is a fixed point on the real line. Evaluate c. Find the density function. Calculate $P[\{x; x > x'\}]$. (Frazer)

6. Show that for the normal distribution

$$F(x) = F_s\left(\frac{x - \mu}{\sigma}\right).$$

7. For the standardized normal distribution prove that for any $x > 0$

$$1 - F_s(x) \equiv \int_x^\infty f_s(y)\mathrm{d}y < \frac{1}{x\sqrt{2\pi}}\,\mathrm{e}^{-\frac{1}{2}x^2}.$$

[Hint: use the fact that $\int_x^\infty y\mathrm{e}^{-\frac{1}{2}y^2}\mathrm{d}y = \mathrm{e}^{-\frac{1}{2}x^2}$]. (Parzen)

8. Prove the results in equations (24) and (25) of the text.

61

9. Evaluate the constant c for the density function

$$f(x) = cx^2 e^{-x^2/\alpha^2} \qquad x > 0,$$
$$= 0 \qquad x < 0,$$

where α is a positive parameter. This distribution is called the *Maxwell distribution*.

10. Evaluate the constant c_{ab} for the density function

$$f(x) = c_{ab}\, x^{a-1}(1-x)^{b-1} \qquad 0 < x < 1,$$
$$= 0 \qquad \text{otherwise,}$$

where a and b are positive integers. This distribution is called the *beta distribution*. Find the distribution function for the case $a = 3$, $b = 2$.

(Frazer)

11. Evaluate the constant c_n for the density function

$$f(x) = c_n\, x^{n-1} e^{-x} \qquad x > 0,$$
$$= 0 \qquad x < 0,$$

where n is a positive integer. This distribution is called the *gamma distribution*, and it can be defined for any positive n, not necessarily integral. Find the distribution function for $n = 1$, and $n = 2$.

12. A continuous distribution has density function $f(x)$ and distribution function $F(x)$. Show that

$$\int_0^\infty [1 - F(x)]\mathrm{d}x = \int_0^\infty yf(y)\mathrm{d}y,$$

and

$$\int_{-\infty}^0 F(x)\mathrm{d}x = -\int_{-\infty}^0 yf(y)\mathrm{d}y$$

assuming that $F(x) \sim 1 - x^{-1-\delta}$ as $x \to \infty$, and $F(x) \sim x^{-1-\delta}$ as $x \to -\infty$, where δ is a small positive number.

CHAPTER SIX

Markov Chains

6.1. Introduction

In this chapter we return to discrete sample spaces and study a more general kind of process than the ones discussed earlier.

We assume that we have a sequence of trials and that the sample space of each trial consists of a finite number of sample points, so that $S = E_1 \cup E_2 \cup ... \cup E_n$. It is assumed that the probability of outcome E_i in any trial is not necessarily independent of the outcomes of previous trials but depends at most upon the outcome of the immediately preceding trial. Thus, we assume that there are numbers p_{ij} which represent the probability of outcome E_j in any particular trial, given that outcome E_i occurred in the preceding trial. In physical applications the outcomes $E_1, E_2, ..., E_n$ are called *states*, and the numbers p_{ij} are called *transition probabilities*. In addition to the numbers p_{ij} we must be given the probability a_k of the outcome E_k in the initial trial. With this information we can calculate probabilities of outcomes in a whole sequence of trials. A process of this kind is called a *Markov chain*.

6.2. Stochastic matrices

In the study of Markov chains a convenient way of presenting the transition probabilities p_{ij} is to form the matrix*

---

* For a detailed treatment of matrices see, for example, P. M. Cohn, *Linear equations*, Ch. 3, Routledge & Kegan Paul, Library of Mathematics.

$$P = \begin{bmatrix} p_{11} & p_{12} & p_{13} & \cdots & p_{1n} \\ p_{21} & p_{22} & p_{23} & \cdots & p_{2n} \\ \cdot & \cdot & \cdot & \cdots & \cdot \\ p_{n1} & p_{n2} & p_{n3} & \cdots & p_{nn} \end{bmatrix} \tag{1}$$

where n is the number of states in the Markov chain. It is clear that P is a square matrix with non-negative elements

$$p_{ij} \geqslant 0 \tag{2}$$

since these numbers represent probabilities. Also, since some outcome follows a particular outcome E_i, it follows that

$$p_{i1} + p_{i2} + \ldots + p_{in} = 1 \ (i = 1, 2, \ldots, n), \tag{3}$$

that is, the sum of the elements in any row is equal to 1. A matrix P whose elements satisfy conditions (2) and (3) is called a *stochastic matrix*.

We may therefore state

Definition. A sequence of trials with possible outcomes $E_1, E_2, \ldots$, initial probability distribution $\{a_k\}$, and transition probabilities defined by a stochastic matrix P is called a *Markov chain*.

Example 1 Consider the stochastic matrix

$$P = \begin{bmatrix} p & q \\ p' & q' \end{bmatrix}, p + q = p' + q' = 1 \tag{4}$$

There are two states E_1 and E_2 in the corresponding Markov chain, E_1 for success and E_2 for failure, say. Then p corresponds to p_{11}, that is, p is the probability of success following success, and similarly q is the probability of failure following success, and so on.

Example 2 Consider a three-state process with stochastic matrix

$$P = \begin{bmatrix} \frac{1}{2} & \frac{1}{2} & 0 \\ 0 & \frac{1}{3} & \frac{2}{3} \\ 1 & 0 & 0 \end{bmatrix} \tag{5}$$

Conditions (2) and (3) are satisfied by this matrix. In (5) the transition probabilities are

$$p_{11} = \tfrac{1}{2}, \qquad p_{12} = \tfrac{1}{2}, \qquad p_{13} = 0,$$
$$p_{21} = 0, \qquad p_{22} = \tfrac{1}{3}, \qquad p_{23} = \tfrac{2}{3}, \tag{6}$$
$$p_{31} = 1, \qquad p_{32} = 0, \qquad p_{33} = 0.$$

Thus, the probability of transition from state E_1 to state E_1 is $p_{11} = \tfrac{1}{2}$, while the probability of transition from state E_1 to state E_3 is $p_{13} = 0$, which means this transition is impossible.

6.3. *r* Step processes

The stochastic matrix P discussed above contains elements p_{ij} which give the probability that the system will move from state E_i to state E_j in *one* step. We now consider the following problem. If the system starts in state E_i with probability $a_i^{(0)}$, what is the probability that it will be in state E_j after r steps? We denote this probability by $a_j^{(r)}$ (read 'a_j for r steps'). Note that this does *not* mean the rth power of a_j. We are interested in this probability for all possible starting states E_i and all possible final states E_j.

To make the discussion simple we consider the case of a two-state process with stochastic matrix

$$P = \begin{bmatrix} p_{11} & p_{12} \\ p_{21} & p_{22} \end{bmatrix}. \tag{7}$$

The probability that the system starts in state E_i is given by the ith component of the vector $\mathbf{a}^{(0)} = (a_1^{(0)}, a_2^{(0)})$. Similarly, the probability that the system will be in state E_j after r steps is given by the jth component of the vector $\mathbf{a}^{(r)} = (a_1^{(r)}, a_2^{(r)})$.

Definition. A row vector $\mathbf{a}$ is called a *probability vector* if it has non-negative components which sum to 1.

It is clear that the vectors $\mathbf{a}^{(0)}$ and $\mathbf{a}^{(r)}$ are probability vectors. Also, each row of a stochastic matrix is a probability vector.

Suppose now that we have taken $(r - 1)$ steps and arrived at state E_1 with probability $a_1^{(r-1)}$, or arrived at state E_2 with

F

probability $a_2^{(r-1)}$. For the next step we can draw the diagram

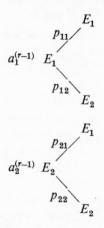

FIG. 6.1. Transition diagram

From the diagram we see that the probability $a_1^{(r)}$ of being in state E_1 after r steps is

$$a_1^{(r)} = a_1^{(r-1)}p_{11} + a_2^{(r-1)}p_{21},$$

and similarly

$$a_2^{(r)} = a_1^{(r-1)}p_{12} + a_2^{(r-1)}p_{22}.$$

Now we can write these in matrix form

$$(a_1^{(r)}, a_2^{(r)}) = (a_1^{(r-1)}, a_2^{(r-1)}) \begin{bmatrix} p_{11} & p_{12} \\ p_{21} & p_{22} \end{bmatrix} \tag{8}$$

or

$$\mathbf{a}^{(r)} = \mathbf{a}^{(r-1)}P. \tag{9}$$

Taking particular values of r in (9) we get

$$\mathbf{a}^{(1)} = \mathbf{a}^{(0)}P,$$

$$\mathbf{a}^{(2)} = \mathbf{a}^{(1)}P = \mathbf{a}^{(0)}P^2,$$

$$\mathbf{a}^{(3)} = \mathbf{a}^{(2)}P = \mathbf{a}^{(0)}P^3, \text{ etc.}$$

In general it can be seen that

$$\mathbf{a}^{(r)} = \mathbf{a}^{(0)}P^r. \tag{10}$$

This means that by multiplying the row vector $\mathbf{a}^{(0)}$ of the initial probabilities by the rth power of the stochastic matrix P, we obtain the row vector $\mathbf{a}^{(r)}$ whose components are the probabilities of being in the various states after r steps.

Equation (10) has been derived for a two-state process. It also holds for an n-state process with n any positive integer.

Example 1 Consider the two-state system with stochastic matrix

$$P = \begin{bmatrix} p_{11} & p_{12} \\ p_{21} & p_{22} \end{bmatrix},$$

and let the number of steps be $r = 2$. To use (10) we therefore require P^2. Now

$$P^2 = \begin{bmatrix} p_{11} & p_{12} \\ p_{21} & p_{22} \end{bmatrix} \begin{bmatrix} p_{11} & p_{12} \\ p_{21} & p_{22} \end{bmatrix}$$

$$= \begin{bmatrix} p_{11}\,p_{11} + p_{12}\,p_{21}, & p_{11}\,p_{12} + p_{12}\,p_{22} \\ p_{21}\,p_{11} + p_{22}\,p_{21}, & p_{21}\,p_{12} + p_{22}\,p_{22} \end{bmatrix} \tag{11}$$

We take two cases.

Case (i). Assume that the system starts off in state E_1 so that

$$\mathbf{a}^{(0)} = (1, 0). \tag{12}$$

Then from (10), (11) and (12) we have

$$\mathbf{a}^{(2)} = \mathbf{a}^{(0)}P^2 = (p_{11}\,p_{11} + p_{12}\,p_{21},\, p_{11}\,p_{12} + p_{12}\,p_{22}). \tag{13}$$

Hence in this case the probability of being in state E_1 after two steps is $a_1^{(2)} = p_{11}\,p_{11} + p_{12}\,p_{21}$, while the probability of being in state E_2 after two steps is $a_2^{(2)} = p_{11}\,p_{12} + p_{12}\,p_{22}$. Note that the components of $\mathbf{a}^{(2)}$ sum to 1, as they should.

Case (ii). Assume that the system starts off in state E_2 so that

$$\mathbf{a}^{(0)} = (0, 1). \tag{14}$$

Then by (10), (11) and (14) we have

$$\mathbf{a}^{(2)} = \mathbf{a}^{(0)}P^2 = (p_{21}\,p_{11} + p_{22}\,p_{21},\, p_{21}\,p_{12} + p_{22}\,p_{22}). \tag{15}$$

So in this case the probability of being in state E_1 after two steps is $a_1^{(2)} = p_{21} p_{11} + p_{22} p_{21}$, while the probability of being in state E_2 after two steps is $a_2^{(2)} = p_{21} p_{12} + p_{22} p_{22}$.

Example 2 Consider the stochastic matrix P given in equation (5) above:

$$P = \begin{bmatrix} \frac{1}{2} & \frac{1}{2} & 0 \\ 0 & \frac{1}{3} & \frac{2}{3} \\ 1 & 0 & 0 \end{bmatrix}$$

By matrix multiplication we find that

$$P^3 = \begin{bmatrix} \dfrac{11}{24} & \dfrac{19}{72} & \dfrac{5}{18} \\ \dfrac{5}{9} & \dfrac{10}{27} & \dfrac{2}{27} \\ \dfrac{1}{4} & \dfrac{5}{12} & \dfrac{1}{3} \end{bmatrix}$$

Hence the probability of reaching state E_2 from state E_1 after three steps is 19/72, and so on.

We see that the rows in P^3 sum to 1. Since the elements in P^3 are all non-negative it follows that P^3 is a *stochastic* matrix.

This result holds in general: if P is a stochastic matrix then any power of P is also a stochastic matrix. To prove this we note that P is stochastic if and only if

$$Pe = e, \tag{16}$$

where e is the column vector all of whose components are unity. Then

$$P^2 e = Pe = e$$

and, in general,

$$P^r e = e \tag{17}$$

for all positive integers r. Hence P^r is stochastic.

Example 3 For *independent* trials, $p_{ij} = a_j$ for all i, and the corresponding stochastic matrix for an n-state system is

$$P = \begin{bmatrix} a_1 & a_2 & a_3 \dots a_n \\ a_1 & a_2 & a_3 \dots a_n \\ . & . & . \dots . \\ a_1 & a_2 & a_3 \dots a_n \end{bmatrix} \tag{18}$$

with $a_1 + a_2 + a_3 + \dots + a_n = 1$.

6.4. Ergodic Markov chains

Consider an n-state Markov chain described by stochastic matrix P with elements p_{ij}. The rth power of P is a stochastic matrix with elements $p_{ij}^{(r)}$, say. If, as r tends to infinity, the elements $p_{ij}^{(r)}$ tend to a limit that depends only on the final state j and not on the initial state i, that is,

$$p_{ij}^{(r)} \to q_j, \tag{19}$$

the Markov chain is said to be *ergodic*. The probabilities q_j are called the *stationary* probabilities for the Markov chain. Another way of stating this is to say that the Markov chain is ergodic if

$$P^r \to Q \tag{20}$$

where Q is a matrix with identical rows, like the matrix in equation (18).

Now $$PP^n = P^nP = P^{n+1},$$

and if Q exists it follows, by letting $n \to \infty$, that

$$PQ = QP = Q. \tag{21}$$

This equation enables us to find Q once its existence is assumed. For example, consider the stochastic matrix

$$P = \begin{bmatrix} \frac{1}{3} & \frac{2}{3} \\ \frac{2}{3} & \frac{1}{3} \end{bmatrix} \tag{22}$$

and assume that Q exists and has the form

$$Q = \begin{bmatrix} q_1 & q_2 \\ q_1 & q_2 \end{bmatrix}, \text{ with } q_1 + q_2 = 1. \tag{23}$$

Then from (21) $Q = QP$ and we have

$$\begin{bmatrix} q_1 & q_2 \\ q_1 & q_2 \end{bmatrix} = \begin{bmatrix} q_1 & q_2 \\ q_1 & q_2 \end{bmatrix} \begin{bmatrix} \tfrac{1}{3} & \tfrac{2}{3} \\ \tfrac{2}{3} & \tfrac{1}{3} \end{bmatrix}$$

$$= \begin{bmatrix} \tfrac{1}{3}q_1 + \tfrac{2}{3}q_2, & \tfrac{2}{3}q_1 + \tfrac{1}{3}q_2 \\ \tfrac{1}{3}q_1 + \tfrac{2}{3}q_2, & \tfrac{2}{3}q_1 + \tfrac{1}{3}q_2 \end{bmatrix}.$$

So $\qquad\qquad q_1 = \tfrac{1}{3}q_1 + \tfrac{2}{3}q_2,$

or $\qquad\qquad q_1 = q_2.$

Now $q_1 + q_2 = 1$, and hence

$$q_1 = \tfrac{1}{2} = q_2. \tag{24}$$

These are the stationary probabilities of the ergodic Markov chain defined by the matrix (22).

6.5. Random walk in one dimension

An interesting example of a Markov chain arises in connection with the n state system which has stochastic matrix

$$P = \begin{bmatrix} 1 & 0 & 0 & 0 & \ldots & 0 & 0 & 0 \\ q & 0 & p & 0 & \ldots & 0 & 0 & 0 \\ 0 & q & 0 & p & \ldots & 0 & 0 & 0 \\ . & . & . & . & \ldots & . & . & . \\ 0 & 0 & 0 & 0 & \ldots & q & 0 & p \\ 0 & 0 & 0 & 0 & \ldots & 0 & 0 & 1 \end{bmatrix} \tag{25}$$

If we imagine that the states $E_1, E_2, \ldots, E_n$ correspond to the points $1, 2, \ldots, n$ on the real line, then (25) may be used to describe the motion of a particle from point to point in one dimension, where each transition occurs with a given probability. In fact the probabilities are

$$p_{i,i+1} = p, \quad p_{i,i-1} = q, \quad (i = 2, 3, ..., n-1),$$

$$p_{11} = 1 = p_{nn},$$

and all the other p_{ij} are zero. Hence, except at the end-points, the particle may jump one unit to the right with probability p, or one unit to the left with probability q. No other jumps are possible. At the end-points we have $p_{11} = 1 = p_{nn}$. This means that once E_1 or E_n is reached, the particle stays there with probability 1. So, if the particle starts at E_i ($i \neq 1, \neq n$), it can move through a set of states until it happens to reach E_1 or E_n and that finishes the process. It is called a *random walk with absorbing barriers.*

We can generalize the above stochastic matrix by considering

$$P = \begin{bmatrix} 1 & 0 & 0 & 0 ... 0 & 0 & 0 \\ (1-\alpha)q & \alpha q & p & 0 ... 0 & 0 & 0 \\ 0 & q & 0 & p ... 0 & 0 & 0 \\ . & . & . & . ... . & . & . \\ 0 & 0 & 0 & 0 ... q & \beta p & (1-\beta)p \\ 0 & 0 & 0 & 0 ... 0 & 0 & 1 \end{bmatrix} \quad (26)$$

where $0 \leqslant \alpha \leqslant 1$ and $0 \leqslant \beta \leqslant 1$. For $\alpha = 0 = \beta$, (26) reduces to (25). When $\alpha = 1 = \beta$, no jump to E_1 or E_n is possible since $p_{i1} = 0 = p_{in}$ for $i = 2, 3 ..., n-1$. Hence if the particle starts at E_i ($i \neq 1, \neq n$) it will move from state to state and *never* reach E_1 or E_n. This is called a *random walk with reflecting barriers.*

These simple models have been used in physics to describe one-dimensional diffusion or Brownian motion, in which a particle is subjected to a large number of molecular collisions which impose on it a random motion.

EXERCISES FOR CHAPTER SIX

1. Find the matrix P^2 for the Markov chain determined by the stochastic matrix

$$P = \begin{bmatrix} \frac{1}{3} & \frac{2}{3} \\ \frac{1}{2} & \frac{1}{2} \end{bmatrix}.$$

2. Find the matrices P^2, P^3 and P^4 for the Markov chains determined by the stochastic matrices

$$P_1 = \begin{bmatrix} 1 & 0 \\ 0 & 1 \end{bmatrix}, \; P_2 = \begin{bmatrix} 0 & 1 \\ 1 & 0 \end{bmatrix}.$$

3. Let P be the stochastic matrix

$$P = \begin{bmatrix} p_{11} & p_{12} \\ p_{21} & p_{22} \end{bmatrix}.$$

Show that any power of P is also a stochastic matrix.

4. A Markov chain contains two states E_1 and E_2 and has stochastic matrix

$$P = \begin{bmatrix} \frac{1}{2} & \frac{1}{2} \\ \frac{1}{3} & \frac{2}{3} \end{bmatrix}.$$

The initial states are selected according to the probability distribution $P[E_1] = \frac{1}{2} = P[E_2]$. Find the probability that the system is in state E_1 after the first transition.

5. With the Markov chain defined in exercise 4, find the probability that the system is in state E_1 after the second transition.

6. Let P be the stochastic matrix

$$P = \begin{bmatrix} 0 & \frac{1}{3} & \frac{2}{3} \\ \frac{2}{3} & 0 & \frac{1}{3} \\ \frac{1}{3} & \frac{2}{3} & 0 \end{bmatrix},$$

and suppose the initial probability distribution is $\mathbf{a}^{(0)} = (\frac{1}{2}, \frac{1}{4}, \frac{1}{4})$. Find the probability distributions $\mathbf{a}^{(1)}$, $\mathbf{a}^{(2)}$, and $\mathbf{a}^{(3)}$.

7. A two-state Markov chain has stochastic matrix

$$P = \begin{bmatrix} 0 & 1 \\ 1 & 0 \end{bmatrix}.$$

What is the probability that after r steps the system is in state E_1 if it started in state E_2? Does this probability become independent of the initial state for large r?

8. Determine whether or not the following matrices describe ergodic Markov chains:

$$P_1 = \begin{bmatrix} \frac{1}{2} & \frac{1}{2} \\ \frac{1}{2} & \frac{1}{2} \end{bmatrix}, \quad P_2 = \begin{bmatrix} \frac{1}{2} & \frac{1}{2} & 0 \\ \frac{1}{2} & \frac{1}{2} & 0 \\ 0 & 0 & 1 \end{bmatrix}, \quad P_3 = \begin{bmatrix} 0 & 1 \\ 1 & 0 \end{bmatrix}.$$

9. Find the stationary probabilities for the ergodic Markov chains whose stochastic matrices are

$$\text{(i)} \quad P = \begin{bmatrix} \frac{2}{3} & \frac{1}{3} \\ \frac{1}{3} & \frac{2}{3} \end{bmatrix}, \quad \text{(ii)} \quad P = \begin{bmatrix} \frac{2}{3} & \frac{1}{3} \\ \frac{2}{3} & \frac{1}{3} \end{bmatrix}.$$

10. Consider a sequence of independent tosses of a coin which has probability p of falling heads. Suppose we say that at time n the state E_1 is observed if the trials numbered $n - 1$ and n resulted in (H, H). Similarly states E_2, E_3, and E_4 correspond to (H, T), (T, H), and (T, T). Thus, for example, a transition from E_1 at time n to E_2 at time $n + 1$ occurs if the results are

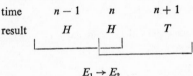

$$E_1 \to E_2$$

Find the stochastic matrix P for this Markov chain. Also, find all the powers of P. (Parzen)

Answers to Exercises

CHAPTER 2. 1. (a) $A' \cap B'$, (b) $(A \cap B') \cup (A' \cap B)$, (c) $A \cap B$, (d) S, (e) $A \cup B$, (f) $A \cap B$, (g) $A' \cap B'$, (h) $(A \cap B)'$, (i) S, (j) $A \cap B'$. **2.** (a) $\{10\}$, (b) $\{1, 2, 3, 7, 8, 9\}$, (c) $\{4, 5, 6\}$, (d) S, (e) $\{1, 2, 3, 4, 5, 6, 7, 8, 9\}$, (f) $\{4, 5, 6\}$, (g) $\{10\}$, (h) $\{1, 2, 3, 7, 8, 9, 10\}$, (i) S, (j) $\{1, 2, 3\}$. **5.** $\frac{1}{6}$, $\frac{23}{36}$, $\frac{1}{3}$. **6.** $\left(\frac{5}{6}\right)^4$. **7.** $\frac{12}{21}$. **8.** $P[A] = \frac{1}{4}$, $P[B] = \frac{1}{2}$, or $P[A] = \frac{1}{2}$, $P[B] = \frac{1}{4}$. **11.** $\frac{1}{3}$, **12.** 3 to 1.

CHAPTER 3. 1. 15, 64. **2.** 26^3, $26^4 + 26^3 + 26^2$. **3.** $2(2^8 - 1)$. **4.** 7670. **5.** 30. **6.** (a) ordered with repetition 49, (b) ordered without repetition 42, (c) unordered with repetition 28, (d) unordered without repetition 21. **8.** (a) $\frac{1}{4}$, (b) $\frac{1}{12}$. **9.** (a) $\frac{1}{n}$, (b) $\frac{1}{n(n-1)}$. **10.** (i) $\frac{1}{20}$, (ii) $\frac{1}{12}$. **11.** For sample space of 12 points (a) $\frac{1}{2}$, (b) $\frac{1}{2}$, (c) $\frac{1}{6}$. **12.** (a) $\frac{7}{8}$, (b) $\frac{2}{3}$.

CHAPTER 4. 1. $\frac{5^2}{6^3}$. **2.** $1 - (0.6)^5 - 2(0.6)^4$. **3.** $\{1 - (0.6)^5 - 2(0.6)^4\}/\{1 - (0.6)^5\}$. **4.** (i) $(1-p)^n$, (ii) $np(1-p)^{n-1}$, (iii) $1 - (1-p)^n - np(1-p)^{n-1}$. **5.** $\frac{5}{16}$. **6.** With 36 sample points, $\frac{8^{n-1}}{9^n}$. **7.** $2^{-2n} \sum_{k=0}^{n} \binom{n}{k}^2 = \binom{2n}{n} 2^{-2n}$. **8.** pq^{n-1}.

CHAPTER 5. 2. $\frac{7}{27}$.

3. $F(x) = 0$ $x < 0$
$\qquad\quad = 3x^2 - 2x^3$ $0 < x < 1$
$\qquad\quad = 1$ $x > 1$.

Answers to exercises

4. $c = \dfrac{1}{\pi}$. $F(x) = \frac{1}{2} + \dfrac{1}{\pi}\tan^{-1}x$. $P[\,\{x; 0 < x < 1\}\,] = \dfrac{1}{4}$.

5. $c = x_0^{\alpha}$.

$$f(x) = 0 \qquad\qquad x < x_0, \quad P[\{x; x > x'\}] = 1 \qquad x' < x_0$$
$$= \alpha x_0^{\alpha} x^{-\alpha-1} \quad x \geqslant x_0. \qquad\qquad\qquad = \left(\dfrac{x_0}{x'}\right)^{\alpha} \quad x' > x_0.$$

9. $c = \dfrac{4}{\sqrt{\pi}}\,\dfrac{1}{\alpha^3}$.

10. $c_{ab} = \dfrac{(a+b-1)!}{(a-1)!(b-1)!}$. $\quad F(x) = 0 \qquad\qquad x < 0$
$$\qquad\qquad\qquad\qquad\qquad\quad = 4x^3 - 3x^4 \quad 0 < x < 1$$
$$\qquad\qquad\qquad\qquad\qquad\quad = 1 \qquad\qquad x > 1.$$

11. $c_n = \dfrac{1}{(n-1)!}$.

For $n = 1, F(x) = 0 \qquad x < 0$
$$\qquad\qquad\quad = 1 - e^{-x} \quad x > 0.$$
For $n = 2, F(x) = 0 \qquad\qquad x < 0$
$$\qquad\qquad\quad = 1 - e^{-x} - xe^{-x} \quad x > 0.$$

CHAPTER 6. 1.
$$\begin{bmatrix} \dfrac{4}{9} & \dfrac{5}{9} \\[2mm] \dfrac{5}{12} & \dfrac{7}{12} \end{bmatrix}.$$

2. $P_1^2 = P_1^3 = P_1^4 = P_1$. $\quad P_2^2 = P_1, P_2^3 = P_2, P_2^4 = P_1$. $\quad$ 4. $\dfrac{5}{12}$. $\quad$ 5. $\dfrac{29}{72}$.

6. $\left(\dfrac{1}{4},\ \dfrac{1}{3},\ \dfrac{5}{12}\right)$, $\left(\dfrac{13}{36},\ \dfrac{13}{36},\ \dfrac{10}{36}\right)$, $\left(\dfrac{12}{36},\ \dfrac{11}{36},\ \dfrac{13}{36}\right)$.

7. 0 if r is even, 1 if r is odd. No. 8. P_1 is ergodic. P_2 and P_3 are not ergodic. 9. (i) $q_1 = \frac{1}{2} = q_2$, (ii) $q_1 = \frac{2}{3}, q_2 = \frac{1}{3}$.

10. $P = \begin{bmatrix} p & q & 0 & 0 \\ 0 & 0 & p & q \\ p & q & 0 & 0 \\ 0 & 0 & p & q \end{bmatrix}$. $\quad P^r = \begin{bmatrix} p^2 & pq & pq & q^2 \\ p^2 & pq & pq & q^2 \\ p^2 & pq & pq & q^2 \\ p^2 & pq & pq & q^2 \end{bmatrix}, r \geqslant 2.$

75

Suggestions for Further Reading

1. Feller, W. *An Introduction to Probability Theory and its Applications, Volume I*, Wiley, New York, 1957.

2. Frazer, D. A. S. *Statistics: An Introduction*, Wiley, New York, 1958.

3. Parzen, E. *Modern Probability Theory and its Applications*, Wiley, New York, 1960.

Index